Multiaccess, Reservations & Queues

Philips Research, Volume 10

Editor-in-Chief
Dr. Frank Toolenaar
Philips Research Laboratories Eindhoven The Netherlands

Scope to the 'Philips Research Book Series'

As one of the largest private sector research establishments in the world, Philips Research is shaping the future with technology inventions that meet peoples' needs and desires in the digital age. While the ultimate user benefits of these inventions end up on the high-street shelves, the often pioneering scientific and technological basis usually remains less visible.

This 'Philips Research Book Series' has been set up as a way for Philips researchers to contribute to the scientific community by publishing their comprehensive results and theories in book form.

Dr. Rick Harwig

Dee Denteneer • Johan S.H. van Leeuwaarden

Multiaccess, Reservations & Queues

 Springer

Authors

Dee Denteneer
Philips Research
HTC 37 (WY5.005)
5656 AE Eindhoven
The Netherlands
dee.denteneer@philips.com

Johan S.H. van Leeuwaarden
Eindhoven University of Technology
and EURANDOM
5600 MB Eindhoven
The Netherlands
j.s.h.v.leeuwaarden@tue.nl

© of the cover illustration by Paul Legeland, The Netherlands

ISBN: 978-3-642-08883-4   e-ISBN: 978-3-540-69317-8

ACM Computing Classification (1998): B.8.1, C.2.1, C.4

© 2008 Springer-Verlag Berlin Heidelberg
Softcover reprint of the hardcover 1st edition 2008

Printed on acid-free paper

9 8 7 6 5 4 3 2 1

springer.com

Foreword

Reservation procedures constitute the core of many popular data transmission protocols. They consist of two steps: A request phase in which a station reserves the communication channel and a transmission phase in which the actual data transmission takes place. Such procedures are often applied in communication networks that are characterised by a shared communication channel with large round-trip times. In this book, we propose queueing models for situations that require a reservation procedure and validate their applicability in the context of cable networks.

We offer two approaches to better understand the performance of these reservation procedures, both based on mathematical modelling. The first, decompositional, approach proposes separate models for the request and the data-transmission phase. In doing so, one ignores the detailed interactions between the two phases of the reservation procedure. Thus, this approach provides insight into the average packet delay but it fails to capture delay variations as required for quality-of-service specifications. To obtain results on higher-order statistics of the delay, we also take a second, integrated, approach based on tandem queueing models with shared service capacity. It is shown that these models can be used to derive accurate approximations for the packet delay in reservation procedures. Moreover, insights obtained from these modelling efforts lead to actual improvements in the data-transmission scheduling. These theoretical claims are supported by simulations of data transmissions in cable networks.

The book proceeds via the study of four key performance models, and modifications to these: contention trees, the repairman model, the bulk service queue, and tandem queues. As such, the relevance of our book is not limited to reservation procedures and cable networks, and performance analysts from a variety of areas may benefit, as all models have found applications in other fields as well.

Acknowledgments. In 2000, Philips Research and the research institute EURANDOM joined forces in the Pelican project to analyse reservation procedures for cable access networks with an approach based on mathematical modelling. This has resulted in a large number of technical reports, patent applications, publications, and two dissertations. In this monograph, we summarise our main findings and theories and thereby complete this exciting task.

This monograph is thus the result of a collaborative effort, and we take this opportunity to thank the various colleagues who have helped us in this undertaking. Onno Boxma was the EURANDOM project leader in the Pelican project and promotor of both DD and JvL; without him neither the project nor the monograph would have seen the light. Sem Borst, Mike Keane, and Jacques Resing acted as co-advisors to our theses; our thanks go out to them for their contributions to the various chapters in this monograph, as well as for their marked influence on the presentation of the material. In particular, Jacques greatly contributed through our many joint projects [22, 23, 51, 111–113].

During the Pelican project, we have also collaborated with Ivo Adan [4, 53], Onno Boxma [22, 23], Christian Gromoll [49], Ewa Hekstra-Nowacka [78, 145], Guido Janssen [50, 87–89], Mike Keane [54], Verus Pronk [55, 78, 145], Ludo Tolhuizen [78, 145], and Erik Winands [4]. Some of the results from these papers have found their way into this monograph.

In addition to those named above, there have been a number of participants in the Pelican project, whose contributions are more implicit. Here, we take the opportunity to thank them for their stimulus, which has helped a great deal in shaping our work: Mark van den Broek, Sebastian Egner, Lerzan Örmeci, Zbigniew Palmowski, Ronald Rietman, Vitali Romanov, Sabine Schlegel, Sai Shankar, and Jaap Wessels. Paul Legeland helped in giving form to the monograph with the drawings that can be found in Chap. 2.

This industrial research project could not have been carried out without the support of the management of Philips Research. Our particular thanks go out to Emile Aarts and Carel-Jan van Driel for their help in getting the Pelican project going. We thank Philips Research and EURANDOM for providing stimulating and enriching environments for both research and collaboration.

Contents

PART I

PROLOGUE

Chapter 1

MULTIACCESS IN CABLE NETWORKS

This book proposes and analyses delay models for communication over a shared channel by means of a reservation procedure. In this introductory chapter, we discuss in general terms the main concepts for multiaccess communication: Contention resolution and reservation procedures. We then turn to cable networks and describe in some detail the current protocols to transmit data. Moreover, we give examples, obtained by complex system simulations, of the specific performance issues that arise in the evaluation of these networks. These examples have strongly motivated our research effort, in which we aim at complementing the simulated results with analytical models and calculations.

1.1   Multiaccess Communication

Multiaccess communication is a well established research topic, see, e.g. Bertsekas and Gallager [14], Hayes [77], or Tanenbaum [160]. The central setting of multiaccess communication is that a number of entities use the same communication channel to transmit messages. This setting will be familiar to most of us from everyday experience. We all use the same channel (the air) for talking. One of the issues associated with multiaccess communication is then also clear from daily life: Speech becomes unintelligible if several people are talking at the same instant.

The examples below serve to illustrate the wide variety of application areas in machine-to-machine communication in which multiaccess communication is relevant.

EXAMPLE 1.1 **Computer networks.** In recent years, computers have become interconnected to form computer networks. This enables them to exchange messages. Frequently, the link between two computers overlaps with similar links between other computers. This is the case in wired and wireless local area networks with either the Ethernet [122] or the IEEE 802.11 standard [106], or in access networks with, e.g. the DOCSIS [58], the IEEE 802.14 [80], or the DVB-DAVIC [59] standard.

Whatever the precise nature of the computer network and its links, transmission of different messages over the same link at the same time usually causes interference which in turn causes message loss. Hence, mechanisms must be implemented to achieve satisfactory link sharing.

A popular way to achieve this, and indeed the method implemented in the protocols mentioned above, is time division: A message is only transmitted successfully if it does not coincide in time with another message transmitted over the same link.

EXAMPLE 1.2 **Radio frequency tagging.** It is likely that in the near future many consumer goods will be labeled with identification tags that can be read automatically from a distance without line-of-sight, see Finkenzeller [67] or Law et al. [107]. Tagging is useful for object tracking, as in luggage handling in airports, or for automatic purchase handling in shops.

Tags are low-cost, passive devices. However, they can take energy from certain radio frequency signals emitted by a tag-reader. The power obtained from this signal allows them to respond, and the response will typically contain an identification of the tagged object. Upon receipt of this response, the tag-reader can retrieve information relevant to the object from a database. This information can then be used for a routing decision, as in luggage handling systems. Alternatively, this information can be a price, so that the purchase can be handled automatically.

If multiple tagged objects respond simultaneously to the activating radio signal, the responses will interfere and the object identifications will be unintelligible to the tag-reader.

EXAMPLE 1.3 **On chip communication.** Systems on Chips (SoCs) are integrated circuits that offer the functionality of a complete system on a single chip; examples of SoCs are single-chip televisions, MPEG encoders, and so on. These SoCs comprise an ever increasing number of basic devices, see, e.g. Jantsch and Tenhunen [?]. These devices form the building blocks that are combined to implement the full functionality of the system.

The devices must be interconnected. Currently, most SoCs have a shared medium architecture for their interconnection network, most often a backplane bus, see, e.g. Benini and de Micheli [13]. In this architecture, there is one communication channel, the bus, that is shared by all the devices. In order to transmit a message, a device must first gain bus mastership, because the bus does not support simultaneously transmitted messages. Hence, bus arbitration mechanisms are necessary when several devices attempt to use the bus simultaneously.

In these three examples, there is contention for the use of the communication channel between the various nodes in the network. Conflicts cause message loss, so that there is a need for some form of conflict resolution in order to guarantee that every node eventually gets hold of the channel and can successfully transmit its message. Ideally, the conflict resolution is organised in such a way that successful message transmission is achieved in the shortest possible time.

How to organise the conflict resolution depends on the context. Again, examples from daily life may illustrate the point. For example, in dinner conversation, it is all right to just start talking when we think of something to say. But in a business meeting it might be the chairman's job to determine who speaks next.

Basic methods to organise access in machine communication resemble these simple examples. Thus in some protocols it is allowed that stations just attempt transmission without consideration of other stations. Stations then continue to transmit their message until it is successfully transmitted. Other protocols dictate a more systematic approach in which stations take turns. Between these extremes there is a whole spectrum of possible protocols. We now give a more systematic account of these access methods.

1.2 Methods for Multiaccess

The standard references on computer networks [14, 77, 160] cover the basic methods for multiaccess: Frequency Division Multiple Access (FDMA), Code Division Multiple Access (CDMA), and Time Division Multiple Access (TDMA). With FDMA, stations are assigned different frequency channels, so that conflicts are avoided as collisions in a frequency band can no longer occur. With CDMA, signals from stations are encoded in such a way that conflicts with messages from other stations are experienced as 'white noise'. Thus, conflicts can be resolved as the noise can be removed using the appropriate decoders. We will be concerned exclusively with TDMA networks, in which simultaneously transmitted messages are always corrupted. Therefore, the channel must be divided over time among the different stations.

There are a number of techniques for implementing TDMA. Hayes [77], Sect. 2.7.4, considers polling, token passing, and random access protocols to achieve TDMA. In polling, a central server addresses all stations in the network in turn. A station that receives a polling message can interrupt this polling cycle to transmit messages. After this transmission, the polling cycle continues. In token passing, there is no central entity that organises the access. Rather, the stations pass a token among each other. This token is a licence to transmit, so that the station in possession of the token can transmit its messages. Upon completion of the transmission, it passes the token on to the next station.

Polling and token passing implement a systematic approach to multiaccess, in which stations are polled, in fixed order, for data transmission. Conflicts are thus avoided. The performance of these systematic approaches depends on the characteristics of the network and the communication needs of the stations. Generally, these techniques are less suitable if there are many stations in the network and if their communication patterns are very bursty. In these circumstances, one prefers random access protocols, see Tsybakov [163].

The best known random access protocol is ALOHA, see Roberts [149]. Using ALOHA, stations transmit their messages immediately, without any coordination with other stations. Therefore, conflicts can arise, and it is necessary that the stations obtain some form of feedback concerning the success of their transmission. This feedback can be obtained through an acknowledgement sent by a receiving station or by a central scheduler. Alternatively, the station can listen to the channel and detect transmission conflicts.

If a transmission error occurs, the station waits a random 'back-off' time and then retransmits its message. This procedure is repeated, possibly with increasing back-off times, until successful transmission. There are many variants of ALOHA that differ as to how this random time is exactly determined. The best of these, slotted stabilised ALOHA, achieves a channel utilisation of $e^{-1} \approx 36\%$ for an open, infinite-population model with Poisson arrivals, see Bertsekas and Gallager [14], Sect. 4.2.3.

In case the stations have the ability to quickly check channel status and interrupt the transmission of their messages, improvements of ALOHA are possible. These improvements are called Carrier Sense Multiple Access (CSMA) or Carrier Sense Multiple Access with Collision Detection (CSMA-CD). Throughputs for these improvements depend on the speed with which transmission errors can be detected, but are much larger than the 36% that can be achieved with slotted stabilised ALOHA. In fact, if transmission errors can be detected infinitely fast, throughputs arbitrarily close to 100% can be achieved via CSMA-CD, which explains the wide-spread popularity of this protocol.

If it is impossible to quickly detect transmission errors, more sophisticated random access protocols are preferable to ALOHA. These algorithms are based on so-called contention trees as introduced independently by Tsybakov and Mikhailov [164] and Capetanakis [30]. Using these techniques, the channel utilisation can be increased to approximately 49%, see Tsybakov and Mikhailov [165] and Mosely and Humblet [127]. We defer a more elaborate introduction of contention trees until Sect. 2.1.

Random access protocols can be used to directly transmit a message. However, they can also be used somewhat differently, as a signal to indicate the intention to transmit a message, i.e. as a request message for data transmission. Once this intention has been successfully transmitted, the message itself can be transmitted without the risk of being corrupted by interrupting messages. These approaches are known as reservation or request-grant procedures. In such procedures, stations request access to the shared medium by sending request messages in contention with other stations. The request messages will be small, relative to the actual message to be transmitted, and run the risk of being lost due to collisions with other requests. Hence, some form of conflict resolution is needed to ensure that a request will eventually get through. However, once the request gets through, the station will, in due time, be granted exclusive access to the channel so that it can transmit its messages without colliding.

Reservation procedures have a clear advantage over transmitting complete messages in contention with other messages: Only the small request messages run the risk of being lost. Therefore, these reservation procedures have the promise of efficient and fast data transmission. Not surprisingly, reservation procedures have been extensively investigated and have found application in many computer communication protocols.

In this book we develop analytical models for studying the delay in reservation procedures. Many aspects of these models are relevant to all protocols that implement conflict resolution. We are, however, motivated in particular by the use of reservation procedures in cable networks that involve contention trees. The models in this book are all developed to deal with specific issues that arise in this context. We now first discuss cable networks and their reservation procedures, and then turn to these modelling issues.

1.3   Data Transfer in Cable Networks

A schematic view of a cable network is given in Fig. 1.1. In order to transmit messages the stations must use the upstream channel, which is shared with the other stations. Messages can only be received via the downstream channel. As stations cannot sense the upstream channel, they cannot communicate with each other directly nor can they check the success of their own transmissions. To provide the essential feedback, cable networks incorporate a central scheduler referred to as the Head-End. The Head-End continuously senses the

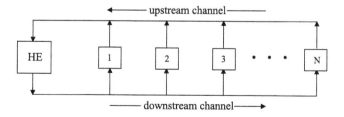

Fig. 1.1.   Schematic view of a cable access network with N stations connected to the Head-End (HE)

upstream channel, observes the success or failure of transmitted messages, and informs the stations about this by broadcasting status reports via the downstream channel.

A system's account of the recent upgrade of cable networks to enable interactive services is given in, e.g. [15, 56, 57]. The multiaccess related details can be found in, e.g. [40, 73, 74, 145, 150]. A number of standard protocols for data transfer in cable networks have emerged and we review the issues that are most relevant to this book. Important protocols are IEEE 802.14 [80], DVB-DAVIC [59], and DOCSIS [58]; DOCSIS is by now the generally accepted standard. These standards describe in great detail the physical elements of the network, the channel medium, the formats of the signals, the conversion of packets into signals and vice versa, the error correction, etc. We largely ignore the physical details of data transmission and concentrate on the medium access control (MAC) layer.

Computer communication is commonly organised as a stack of layers, the OSI stack, see, e.g. [14], Sect. 1.3.2. Each layer in this stack offers specific functionalities to the above layers that are needed to establish a connection. In this system view, the data link layer is situated between the physical layer and the network layer. The MAC sublayer is considered as lower part of the data link control layer and regulates the multiaccess channel.

The protocols describe a variety of methods for data transmission, and differ as to which methods are allowed. Generally, it is possible to transmit data both directly and via the reservation procedure. We will concentrate on the analysis of the reservation procedure and consider the following situation described in, e.g. [59, 80]. Stations request access to the upstream channel for data transmission by transmitting a request message. These requests are in contention with the requests from other stations and are transmitted in time slots that have been specifically designated by the Head-End as request slots. As requests can collide with other requests, some form of conflict resolution must be implemented in order to guarantee that every request will eventually get through to the Head-End. We will focus on the situation in which conflicts between request messages are resolved by means of contention trees. These are introduced in Sect. 2.1 and are extensively analysed in Part II of this monograph. After a successful request, data transfer follows in reserved 'data-transmission' slots, not in contention with other stations. These transmission slots are assigned by the Head-End by means of a message broadcast via the downstream channel.

Stations communicate via an exchange of quanta of information called packets. Such a packet can be a request packet or a data packet. Usually, a request packet is smaller than a data packet, and we consider the specific situation in which it is three times as small as a data packet. The upstream channel is time slotted. Each time slot can contain exactly one data packet. Alternatively, three request packets can be fitted into one time slot.

This notion of a data packet as the basic unity of communication allows for a more precise description of the reservation procedure described above. The rule in this reservation procedure is that a station can make a request for data transmission as soon as it has data to transmit. The request message contains a field, called the request size, to indicate the number of data packets for which transmission is requested. The request size equals the number of data packets actually ready for transmission at the instant of the request attempt. Request messages may be lost due to a conflict with other requests. In this case another request message must be sent. This new request message is identical to its predecessor, except for the value of the request size field: The request size may be increased if the number of packets ready for transmission has increased since the previous request attempt. This property is called 'partially gated' in [14], Sect. 3.5.2.

Whether a time slot will be used for data or for requests is decided by the Head-End, which periodically broadcasts the use of the next sequence of time slots to all stations in the network. If a time slot is to be used for data, the Head-End will also assign it to one of the stations, which can then use it, contention free, to transmit data. If a time slot is to be used for requests, the Head-End can restrict the use of this request slot to subsets of stations satisfying various criteria. The decision whether to use a slot as either a request or a data-transmission slot is usually taken simultaneously for groups of 18 consecutive slots, called frames. The duration of such a frame is approximately 3 ms. The Head-End employs a scheduling strategy to determine the frame layout and in particular the number of request slots and the number of data-transmission slots. After the layout has been determined, it is broadcast to all the stations in the network.

It is another important characteristic of cable networks that these scheduling decisions concerning the frame layout must be taken well before the actual start of the frame to which they correspond. There are a number of reasons for this. Firstly, cable networks are large and the number of stations connected to one Head-End is typically in the order of 100–1,000. As a consequence, such networks can cover a wide area, and the distance between stations and Head-End can be quite substantial. Hence, there is a non-negligible propagation delay involved in transmitting a message from a Head-End to a station and vice versa. Secondly, messages in cable networks are interleaved. This means that they are spread out over time, so as to make the error correction more effective against burst errors. However, this also increases the length of the message: The time between the start of a message and its end. Thirdly, it requires processing time at the Head-End to calculate the schedules and at the station to process and interpret the schedules.

The length of the scheduling message (including interleaving) plus the propagation times plus the processing times at both Head-End and station is non-negligible and is at least about equal to the frame duration. Depending on the implementation details, it can be larger, and information from the current frame can only be used some frames later. Therefore, scheduling decisions must be taken well in advance. In particular, there is a delay of at least some frames before the data transfer corresponding to a request can actually be carried out.

In the remainder of this book, this delay will often be referred to as the round-trip time. It is especially relevant for the model considered in Chap. 8.

1.4   Performance Analysis of Cable Networks

A common, simplifying, attempt to understand the performance achieved in multiaccess systems is to view the shared channel as a kind of central 'server' that has to serve a stream of incoming messages, see, e.g. [14], Sect. 4.1. Such models are known as open models, as the packets in the arrival stream are not associated with specific originating stations. Using this approach, the analysis of the delay in cable networks can then proceed along classical lines.

However, the discussion in Sect. 1.3 has emphasised a number of characteristics that show that such a view has its limitations for the performance analysis of cable networks. These limitations emerge mainly as a consequence of the reservation procedure. Firstly, there are at most a finite number of stations that can enter the contention procedure to transmit a request message. Thus, one can expect that there is a finite-population effect that will show up in medium to high load conditions, which is not captured by open models. Secondly, the server must be shared by the reservation process and the data-transmission process and the optimal tuning of these processes is complicated by the propagation delay. The determination of the bandwidth allocated to either process requires a scheduling policy, which does not have a counterpart in the simplifying model described above.

As there are hardly any accurate models to assess the performance of cable networks, their evaluation is usually carried out by means of simulations. We have drawn in particular upon the simulation platform described in Kwaaitaal [104] and Pronk and De Jong [144]. This platform was used to investigate the performance of cable networks in [78, 145]. In the simulations, stations generate data packets according to a Poisson process. These are transmitted over the cable network via a reservation procedure using contention trees. Performance figures are recorded concerning queue sizes and packet delay, yielding important insights. Moreover, it is possible to compare various slot scheduling algorithms that are used in the Head-End.

We now review some of these simulation results. They demonstrate that the finite-population effect and the scheduling effect do indeed manifest themselves as major issues in the performance analysis of cable networks. This sets the main objective for this book: To formulate and analyse performance models that incorporate the finite-population effect and scheduling policies. These models will then enable us to complement the simulations with analytical calculations.

1.4.1 A Finite-Population Effect

In a first experiment, we compare the mean packet delay in three different scenarios. In each scenario the stations generate data, identically and independently, according to a Poisson process, but the scenarios differ in that the number of stations in the network is varied. If we would use standard open queueing models to predict the average delay, then there would be little difference between the three scenarios: As the superposition of a number of Poisson processes is again a Poisson process, only the total traffic volume matters here.

Performance curves for this experiment are presented in Fig. 1.2 for the cases that there are 50, 100, and 200 stations in the network. These curves display the average delay of a transmitted packet against the total traffic intensity in each of the three scenarios. The performance curves for the three scenarios differ considerably, and show that standard open models are not directly applicable in this situation. The simulation results thus point to a finite-population effect: The number of stations in the network strongly affects packet delay.

This finite-population effect is observed in many system simulations of cable networks. In Pronk et al. [145], Sect. 4.2, it is shown that this finite-population effect also shows up in cable networks with other data-transfer modes not considered here. Moreover, Golmie et al. [74], Fig. 17, report a similar effect.

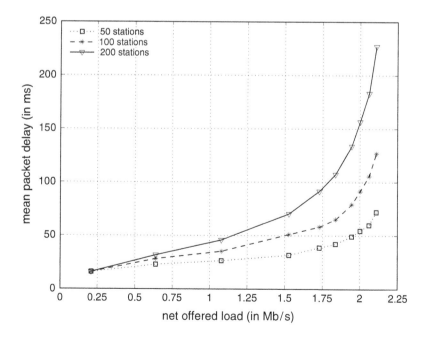

Fig. 1.2. Delay vs. load curves for three different network scenarios, in which networks have different numbers of stations

There is another salient feature in Fig. 1.2, extensively discussed in Denteneer and Pronk [55], which concerns the capacity of the network. As a preliminary fact, we mention that the capacity of the contention tree is approximately $\log(3)$, see Sect. 3.3.1. This means that, on average, $\log(3)$ requests can be processed during one time slot consisting of three request slots. The maximum total traffic intensity in the simulations is approximately equal to 2 Mbps, which corresponds to a traffic intensity of 16 data packets per frame, for frames of 18 slots as described in Sect. 1.3.

If the open queueing model is adopted as a model for data transfer in a cable network, then it is easy to see that the transfer of each packet requires on average $(1+1/\log(3))$ time slots: $1/\log(3)$ for the request and 1 for the packet transfer itself, cf. [14], (4.59). This would limit the capacity of the network to $18/(1 + 1/\log(3)) \approx 9.5$ data packets per frame. However, this upper bound on the capacity is well below the maximum value of 16 packets per frame, mentioned above. This phenomenon can be explained if we consider a closed queueing system. In this case packets originate from a particular station, and it is well possible that stations ask for the transmission of more than one data packet in one request. This further underlines that the models should reflect that there are only a finite number of stations active in the network.

1.4.2 A Scheduling Effect

A second experiment is concerned with the scheduling of time slots as either request slots or data-transmission slots. Here, we fixed a network scenario and considered three different schedules to designate the use of the time slots in a frame. In each of these schedules, scheduling decisions are frame-based. The schedules originate from Pronk et al. [145] and each schedule guarantees a different minimum number of time slots per frame for the request procedure. These slots are devoted to handling requests in contention. The other time slots in a frame are used for data transmission. If there are not enough data packets to use the whole frame, the remaining slots are again used for handling requests in contention.

In Fig. 1.3, we have displayed the simulated performance curves for three distinct schedules. In the simulations, the number of slots guaranteed for the request process equalled 3, 6 and 9, respectively, but the actual values are not important for the discussion. Figure 1.3 shows that the different schedules result in very similar performance at low loads. However, at high loads, the performance curves deviate considerably. Hence, it is apparent that there is a scheduling effect. Moreover, no schedule dominates the others. Rather, each schedule is best in a selected subinterval of the traffic load.

This scheduling effect is not specific to the details of our simulations and has also been observed in other simulations of cable networks. For instance, in Sala et al. [150] system simulations are carried out using ALOHA for requests.

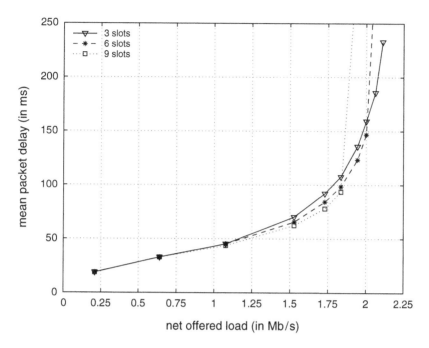

Fig. 1.3. Delay vs. load curves for three different ways to organise the contention process

First, they tested a greedy scheduling strategy, in which slots are only given to the request process if there are no more data to transmit, i.e. if the data queue is empty. It was observed that this strategy resulted in cyclic behaviour of the data queue. Next, it was established experimentally that performance improves if a certain minimum number of slots is guaranteed to requests. Also Golmie et al. [73] consider various schedules for bandwidth allocation to either requests or data transmission. Their results are similar and further stress the relevance of these schedules.

1.5 Outline

Both the finite-population effect and the scheduling effect are prominent features of delay in cable networks as shown by the simulation results reviewed in Sect. 1.4. However, neither has been captured in the performance models for cable networks that have been proposed to date. In this book, we attempt to fill the gap and describe models that account for these effects.

Another way to view this book is suggested by the title and generalises the scope from cable networks to reservation procedures and queueing models. Thus, this book is about the delay analysis of the reservation procedure with contention trees. Our interest is focused on the finite population case and a communication channel with substantial round-trip times. We are now in a position to give an outline of this book.

Part I: Prologue

In Part I, we review the basic notions of multiaccess, cable networks, as well as the various key models that we build upon in the remainder of the book. In fact, there are four fundamental points of departure for the models in this book: Contention trees, the repairman model, the bulk service queue, and tandem queues. As such, the relevance of this book is not limited to cable networks as all models have found (or might find) application in other domains. The key models are reviewed in Chap. 2.

Part II: Contention Trees

The first basic notion is the Capetanakis-Tsybakov contention tree introduced in Capetanakis [30] and Tsybakov and Mikhailov [164]. Algorithms based on contention trees are among the best known algorithms for conflict resolution; we defer a more detailed description of contention trees until Sect. 2.1, where we also give a brief introduction to the contention tree literature.

Part II presents our own contributions to the analysis of contention trees. In Chap. 3, we give a mathematical account of the contention trees and present two formal ways to describe their evolutions. Both ways are based on the complete m-ary tree, also considered in Capetanakis [30] and Kaplan and Gulko [93]. These models are used to analyse a number of basic properties of contention trees. In Sect. 3.3.1 we consider statistics associated with the length of the contention tree, and in Sect. 3.3.3 we review a modification of the standard algorithm, named skipped level trees, which improves the standard contention tree algorithm with respect to the time needed to resolve all conflicts.

We consider the success instants of a contention tree in Sect. 3.3.2. The success instant measures the time from the start of the contention tree until a successful conflict resolution and is thus related to the delay when using contention trees. We use the models to give an expression for the marginal distribution of the success instant, see Theorem 3.3, and to prove that this expression is asymptotically exact when the number of contenders tends to infinity, see Theorem 3.4.

In Chap. 4, we extend the discussion to sequences of contention trees as they are used in practice in dynamic environments. In these environments, stations become inactive after having successfully transmitted their requests. However, as the contention tree is being processed, inactive stations can become active again when they have data to transmit. Hence, the basic contention tree algorithm must be complemented with a 'channel access protocol' or 'first transmission rule': A set of rules to regulate the behaviour of newcomers. In Sect. 4.2, we review the standard channel access protocols: Free access and blocked access. Section 4.2.1 describes an improvement of these standard protocols referred to as scheduled access. The scheduled access protocol improves the basic access protocols both in terms of capacity and delay properties.

We then turn to the request delay, defined as the delay experienced by requests made in contention via a contention tree. The complicated distributions of quantities that depend on the contention tree algorithm preclude a complete analytical treatment. However, the properties of the contention trees reviewed in Chap. 3 suggest that queueing approximations are appropriate for contention trees when used in dynamic environments. Moreover, the finite-population effect suggests that closed queueing models are appropriate.

The remainder of Chap. 4 is devoted to this theme and investigates closed queueing systems to approximate the request delay for the three contention tree protocols. It discusses a number of modifications of the so-called repairman model and gives an analysis of the associated sojourn times, either exactly or heuristically. The proposed modifications pertain to the service discipline. Next to the First Come First Served (FCFS) service discipline, as in the standard version of the repairman model, we also consider Random Order of Service (ROS), Gated Random Order of Service (GROS), and Gated Partial Random Order of Service (GPROS).

These modifications are all motivated by the various channel access protocols that complement the basic contention tree algorithm. We will argue that ROS is appropriate for contention trees with free access, that GROS is appropriate for contention trees with blocked access, and that GPROS is appropriate for contention trees with scheduled access. These observations are supported by simulations, in which we show that the analytical expressions for the first moments of the sojourn time in these repairman models give excellent approximations to the corresponding moments of the request delay for contention trees obtained by simulation.

The repairman model with GROS does not belong to the class of product-form networks, see [10], and we only give a heuristic analysis of its delay properties in Chap. 4. Motivated by the fact that cable networks are usually large, we set out for a more precise mathematical analysis in Chap. 5 using asymptotic analysis. This analysis is utilised to establish the limiting distribution of the request delay, see Theorem 5.2.

Part III: The Bulk Service Queue

In Part III of this book, we turn to the data-transmission delay: The delay due to the queueing of packets at the central scheduler, once a request has been successfully transmitted. We model this part of the reservation procedure by means of the bulk service queue. This is a discrete-time queueing model in which a central server periodically serves a fixed number of customers that arrive according to some random process.

A first introduction to the bulk service queue is given in Sect. 2.3. In Chap. 6 we give an extensive historical introduction of this queueing model, including the main techniques to solve for the stationary distribution. In our treatment, we give the rigorous mathematics, and provide guidelines for the numerical work.

The bulk service queue is introduced in particular to provide a mathematical foundation to analyse the scheduling effect, described in Sect. 1.4.2. The classical bulk service queue can be used to model a simple schedule, in which a fixed fraction of each frame is devoted to the request process. This is a wasteful schedule, as it potentially leaves capacity unused. In Chap. 7, we formulate a flexible-boundary model to represent a better scheduling policy. This schedule was also used in Sect. 1.4.2, and guarantees a fixed minimum amount of slots per frame for the request process. We show that this model can be solved by means of the classical techniques from Chap. 6. However, it fails to accurately capture the scheduling effect, as it does not account for the round-trip times on the communication channel.

Motivated by this shortcoming, we propose to modify the classical bulk service queue by including the delay as another additional element. Due to this delay, arriving packets can only be 'served' after some delay period. We give a thorough motivation for the delayed model in Sect. 2.3 and its detailed analysis is reported in Chap. 8. Although the delayed bulk service queue defines a higher-dimensional Markov chain, we are still able to solve for the stationary distribution of the queue, but the solution hinges on complicated numerical procedures. However, we are able to obtain better insight in the expected queue size by exploiting a method of Kingman [95] that leads to an insightful expression for the expected data-queue size as a function of moments of the arrival distribution, a term related to the idle time, and an auto-correlation term. We then bound the term related to the idle time, and apply a heuristic argument to approximate the correlation term, so that we are able to approximate the expected data-queue size of the delayed bulk service queue.

The approximating bounds are complemented with simulations to establish some interesting properties of the expected data-queue size. Firstly, the expected data-queue size is increasing with the delay. Secondly, and rather remarkably, the expected data-queue size is not monotonic in the traffic intensity, so that a larger traffic volume may actually result in a substantially lower expected data-queue size. Most importantly, however, the delayed bulk service queue aids in the formulation of good scheduling policies to split the bandwidth between the request and data-transmission processes. One such scheduling policy is obtained by a minimisation of the expected queue size with respect to the amount of capacity guaranteed to the request process. A second, more adaptive, schedule is derived by a minimisation of the burstiness of the arrival process, which can be controlled through the request process. The latter property makes the delayed bulk service queue suitable as a model for delay in cable networks, see Sect. 1.4.2.

Part IV: Tandem Queues with Shared Service Capacity

The approaches from Part II and III enable a decompositional approach to analyse the performance of reservation procedures, with separate models for the reservation phase and the data-transmission phase. For an integrated approach, we introduce tandem queues. In the tandem queue, users arrive at a first server, possibly queueing up if the server is busy. Upon service completion at this first server, they move on to a second server, again queueing in front of this server if this second server is busy. After being served at the second server, the users leave the system. In the classical tandem queue, the servers work independently of each other with a fixed service rate. We deviate from this classical model in that the servers in our model share a common service capacity, and this necessitates a service discipline that allocates the service capacity to the servers.

In Chap. 9 we give a detailed treatment of the two-stage tandem queue with shared service capacity. In order to do so, we need to make some rather restrictive assumptions. We assume that users arrive to the request queue according to a Poisson process, and that they require exponentially distributed service times at both queues. Under these assumptions, the tandem queue with shared service capacity gives rise to a two-dimensional Markov process that can be analysed using the theory of *boundary value problems*. A pioneering study of this type of Markov processes is the one of Malyshev [115], whose technique was introduced to queueing theory by Fayolle and Iasnogorodski [61]. We show that the problem of finding the generating function of the joint stationary queue length distribution can be reduced to a Riemann–Hilbert boundary value problem.

Starting from the solution of the boundary value problem, we consider the issues that arise when calculating performance measures like the mean queue size and the fraction of time a queue is empty.

In Chap. 10 we present a more general model of a *two-station network with shared service capacity*. After receiving service at a server, a user either joins the queue of the same server, joins the queue of the other server, or leaves the system, each with a given probability. Users require exponential service times at each server. This general network model covers both the model of Fayolle and Iasnogorodski [61] and the two-stage tandem queue with shared service capacity as special cases. We show that the general model can also be solved using the theory of boundary value problems.

Part V: Epilogue

In Chap. 11, we return to our point of departure and reconsider the reservation procedure in a cable network that motivated our work. In particular, we focus on the delay in cable networks and use the models and their analyses to approximate the total average packet delay. We compare this approximation with the delay figures that are obtained in system simulations of cable networks. Moreover, we use the approximation to quantify the effects of suggestions for scheduling made in this book. We conclude the book with a brief review of open questions for further research.

1.6   Selected Bibliography

As acknowledged in our foreword, this book is the outgrowth of a project on the mathematical analysis of access delay in cable networks. As such, it is based on, and summarises, a number of publications, technical reports, patent filings, and two dissertations. The dissertations are Denteneer [47] and Van Leeuwaarden [109]. We now turn to a brief review of the publications and technical reports.

In Chap. 1, we have built upon the simulation approach described in Pronk and de Jong [144], Pronk et al. [145], and Hekstra-Nowacka et al. [78]. The observations on the finite-population effect stem from Denteneer and Pronk [55]. The material in Chap. 3 is based on Denteneer and Keane [54]. Chapter 4 is based on Boxma et al. [22, 23]. The sections on the gated service discipline have appeared earlier as Denteneer [44, 45]. Scheduled access has been extensively described in Denteneer [46]. The material in Chap. 5 is based on Denteneer and Gromoll [49].

Chapter 6, on the bulk service queue, is taken from Van Leeuwaarden [108]. Theorem 6.1 and the appendix of Chap. 6 are based on joint work with Adan and Winands, see [4]. The first modification of the classical bulk service queue, which couples the arrival process to the queue size, stems from Van Leeuwaarden et al. [111]. The model and the main results from Chap. 8 are taken from Denteneer and Van Leeuwaarden [52] and [50, 51, 53].

Chapters 9 and 10 are based on Van Leeuwaarden and Resing [112, 113]. The material in Chap. 11 stems from Denteneer [48].

Chapter 2

KEY MODELS

This book is about the study of a reservation procedure on a communication channel with a large round-trip time. The study is carried out via the mathematical analysis of a number of key performance models: The contention tree, the repairman model, the bulk service queue, and the tandem queue with shared service capacity. In Chap. 1, we have briefly touched upon the four key models that form the points of departure for this monograph. In this chapter, we give a more extensive introduction to these models and motivate their use within the context of a reservation procedure. Additionally, we give a brief introduction to the vast literature available and highlight some of the original contributions made in this book.

2.1 Contention Trees

Contention trees constitute a popular class of techniques to provide access to a shared resource. They are used on shared, time-slotted, communication channels on which each transmission results in either a perfect transmission or a collision. In case of a collision, all messages involved in the collision are completely lost and must be retransmitted at some later stage. Moreover, all transmitting stations have the ability to detect conflicts, either by listening to their own transmissions or via feedback from some central scheduler. Usually, it is assumed that the feedback is immediate. In [59, 80], contention trees were proposed as a means to resolve the conflicts among the colliding request messages of the reservation procedure as used in cable networks.

Contention trees were introduced independently by Capetanakis [29] and Tsybakov and Mikhailov [164]. Massey [119] and Bertsekas and Gallager [14] both are excellent introductions to the subject. Further mandatory reading on the subject is Volume IT-31, No. 2 of the IEEE Transactions on Information

Theory, which was completely devoted to the mathematical analysis of contention trees. Another noteworthy reference is Janssen and de Jong [86].

The contention tree is initialised when a group of contenders collide, that is, when multiple contenders transmit a message in the same slot. This group of contenders is then, recursively, split by dividing the contenders over the slots in the child node of the slot in which the collision occurred. The division of contenders over the child slots is usually achieved by random choice and each colliding station randomly selects one of the slots in the child node to retransmit its message. Alternatively, identifying properties of the colliding stations, such as identification numbers, can be used to carry out the split. The splitting continues until all conflicts have been successfully resolved. This is achieved as soon as all branches in the tree terminate in leaves that are either empty or have a successful transmission.

This algorithm can be graphically depicted as a tree, or a cactus, as its recursive splitting resembles the branching in either plant. This is illustrated in Fig. 2.1. Note that each branch of the cactus is split into three subbranches. The contention trees employed in cable networks share this ternary character, as each non-terminal slot has a child node with three slots.

The use of such a random access technique raises several important questions. Firstly, one is interested in the efficiency of this technique. How long does it take to resolve all the conflicts, given that there is a certain number of stations initially involved in the conflict? Clearly, the faster the better, as this reduces the time needed for the request phase and we will review a number of variations on the basic contention tree that were specifically developed to improve its efficiency. Secondly, one is interested in the delay caused by using a contention tree, defined as the time elapsed between the initial conflict and the successful reservation of some arbitrary station.

In the original contention tree, conflicts are resolved by recursively splitting the group of contending stations into two subgroups. This splitting is organised as follows. Each of the stations involved in the conflict randomly chooses one out of two subgroups, labeled 'left' and 'right', say. The subgroups then await their turn for retransmission, as indicated by the retransmission rule. Usually, subgroup 'left' can retransmit immediately, whereas subgroup 'right' has to wait until all of the conflicts in the corresponding subgroup 'left' have been resolved. This process of splitting and retransmitting then continues until all messages have been transmitted successfully.

This procedure can be depicted graphically by means of a cactus, see Fig. 2.1, but Fig. 2.2 gives a more common representation. It shows how a tree is formed when an initial group of stations is involved in a conflict, indicated by the 'c' for collision, in the root of the tree. After each split, a subgroup is formed. This subgroup can be empty, resulting in an empty time slot indicated by 'e', it can consist of 1 station, resulting in a successful transmission

Fig. 2.1. The recursive splitting resembles the branching of a tree, or a cactus

indicated by '1', or it can consist of more stations so that another collision occurs, indicated by 'c'. In the latter case, this subgroup is split and the procedure is repeated.

Instead of visualising the splitting procedure as a tree, one can describe the procedure by means of a stack: After each collision, two subgroups are pushed onto the stack, and one subgroup is popped from the stack at the retransmission instant. Therefore, this algorithm is also known as the 'stack' algorithm. Note that the procedure can be implemented at each station by means of a simple counter. Stations that transmit their packet for the first time and that

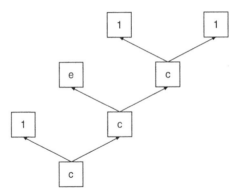

Fig. 2.2. Basic contention tree: Slots of the tree with a collision, i.e. an entry 'c', are recursively split until all slots are empty 'e' or have a successful transmission '1'

experience a collision initialise their counter to 0 or 1, randomly chosen. After this initial collision, stations with a counter equal to 0 may retransmit. Stations not involved in the retransmission add 1 to their counter if they observe a collision and subtract 1 from their counter after an idle slot or a successful transmission.

As an alternative to the stack (or depth-first) order, one can process the collisions in the tree in breadth-first order. In this case, we traverse the tree by layer, see Chap. 3 for a formal definition. This can be advantageous on a transmission channel that is characterised by large round-trip times.

An improvement of this splitting algorithm is obtained by varying the number of subgroups after each collision, see [120]. Another improvement is based on the following simple observation: If a 'left' subgroup is empty, then the 'right' subgroup will certainly consist of more than one station. Consequently, the retransmissions by this right group will necessarily result in a collision. Hence, a more effective conflict resolution can be achieved by skipping this retransmission and splitting this right subgroup before the retransmissions. This observation is due to Massey [119] and leads to the modified tree algorithm.

The discussion above has not touched upon the issue of what to do with newcomers, i.e. stations that need to transmit data but are not participating in the tree algorithm. There are three well-known 'channel access protocols' that regulate the access of newcomers: Free access, blocked access, and windowed access. With free access, newcomers can participate in an ongoing tree, and join the subgroup at the top of the stack. Thus, they can transmit at the first possible transmission instant. In blocked access, newcomers must await the completion of the current tree, and can only enter the contention process in the root node of a new tree. Finally, windowed access, see Gallager [72], is a form of blocked access in which only a limited set of newcomers is allowed to participate in the root node of a tree. Often, this set is defined via a time

Table 2.1. Comparison of (approximate) maximum stable throughput (MST), of various tree algorithms formed by a channel access protocol (CAP), and tree algorithm (TA)

CAP	TA	MST
Free	Basic	0.40
Blocked	Basic	0.36
Blocked	Modified	0.43
Windowed	Modified	0.487
Windowed	Basic	0.43

window, and only newcomers that have become active during this window are allowed to participate in the new tree. The combination of windowed access and the modified tree algorithm leads to the FCFS 0.487 algorithm, see Polyzos and Molle [141].

Combining one of the two tree algorithms with one of the channel access protocols results in a number of conflict resolution algorithms. These differ as to the efficiency with which conflicts are resolved. One way to express this difference is via the maximum stable throughput, which is defined as the maximum arrival rate (for a Poisson arrival process) that can still be handled by the algorithm. In Table 2.1, we have listed the maximum stable throughputs of a number of algorithms. The results in Table 2.1 show that the combination of the modified tree algorithm with windowed access, with appropriate window size, is the most efficient algorithm. Its delay analysis can be found in [141].

Yet, there are difficulties with the use of the modified algorithm in the context of cable networks. The first difficulty is general and shows up in every application of this algorithm in which the channel status is observed with error. Assume that an idle slot is mistaken for a collision. In this case, the optimisation in the splitting rule, which leaves out 'guaranteed' collisions, will lead to indefinite splitting. A second difficulty is more specific to cable networks, as the optimisation is more difficult to implement on a channel with large propagation delays.

Part II of this monograph is devoted to contention trees. We stick to the basic tree algorithm, albeit that we consider general m-ary trees, in which subgroups are split in m rather than two subgroups. Moreover, following Capetanakis [29], we will allow the number of subgroups in the initial split to be different from m. This can be considered as a combination of windowed access and the basic tree algorithm.

There are a number of original contributions in this monograph concerning access delay due to the tree algorithms. In Chap. 3, we study the delay due to the basic blocked contention tree, when the collisions in the tree are processed in breadth-first order. Another innovation is the use of the sojourn time in the repairman model to approximate the access delay when using contention trees, see Chap. 4.

2.2   The Repairman Model

The repairman model, see [64, 159], is one of the key performance models for closed queueing networks, i.e. networks with a finite population of customers. The repairman model is also known as the computer terminal model, see Kleinrock [99], or as the time-sharing system, see Bertsekas and Gallager [14]. The basic model is illustrated in Fig. 2.3. Depicted is a population of ants

Fig. 2.3. Basic repairman model

searching for food. A soon as an ant has found something to eat, it joins a queue in front of the ant hill. On entrance to the hill, the ant stores its catch and resumes its search for food.

A more traditional description of the model is in terms of a repairman and machines (see Fig. 4.1 in Chap. 4). There are N machines working in parallel. After a working period a machine breaks down and joins the repair queue. At the repair facility, a single repairman repairs the machines according to some service discipline. Once repaired, a machine starts working again.

In this book, the repairman model is used to represent the request procedure. The translation is as follows. The working machines stand for the inactive stations that are not participating in the request procedure. The machines in the repair queue correspond to the machines that are active and participating in the request procedure. Stations that are inactive may become active at a random instant. In the parlance of the repairman model, stations break down and then join the repair queue.

In the classical formulation of the model by Palm, see Feller [64], there are N machines and K repairmen. The machines break down after a working period. After this, they are served by one of the repairmen, and, once repaired, the machine starts working again. If all repairmen are busy, the machines queue up at the service facility until one of the repairmen becomes available.

One is typically interested in the fraction of time that all machines are working properly. A related issue is the optimisation of the cost of operation of the machines. In particular, if costs can be assigned both to the unavailability of the machines and to the hiring of repairmen, one can choose the number of repairmen so that some sort of economic optimum is achieved. Another key finding relates to the economies-of-scale: In specific situations, it is more efficient to operate one system with N machines and K repairmen than to operate K systems with each N/K machines and 1 repairman.

In later developments, the model has also received wide-spread popularity as the computer terminal or the time sharing model, see, e.g. [14, 97, 99, 151]. In this version of the model the machines have been replaced by users at a computer terminal and the repairmen are replaced by one central computer which is shared by all users. After some thinking period the user submits a job to this central computer. There the job is served. After service completion, the job is returned to the user, who starts thinking again. Our use of the repairman model is more akin to this computer application: Users are replaced by stations contending to submit a request and the shared computer is replaced by a shared medium to transmit the requests. We focus our investigations on the sojourn time of a job at the central facility, which will then provide a measure for the request delay, defined as the time needed to successfully transmit a request.

When assuming that both the working times and the service times are exponentially distributed, one can solve for the stationary distribution of the number of working machines, which is a truncated Poisson distribution. The number of working machines in the system behaves as the number of occupied trunks in the Erlang loss system, see, e.g. Takács [159], Chap. 5, and the repairman model shares the well known insensitivity property of the Erlang loss system, which states that the distribution of the number of occupied trunks depends on the distribution of the holding times only through the mean holding time. Consequently, in the repairman model, the distributions of the number of working machines and the number of machines in repair only depend on the distribution of the working times via the mean working time. In [159] the system is analysed with exponential working times and general service times.

The discussion so far has ignored the order in which the machines are repaired by the repairman. It is readily argued that the stationary distribution of the number of working machines does not depend on this service discipline, as long as the discipline is work conserving and does not pay attention to the actual service times. However, this is not the case for the sojourn times at the repair facility. In case of the First Come First Served service discipline and exponential working and service distributions, the sojourn time distribution is easily obtained from the distribution of the number of working machines and the arrival theorem, see, e.g. Kobayashi [101] and Chap. 4. In case of the First Come First Served service discipline, exponential working times, and general service distribution, the sojourn times are studied in [159]. In Mitra [126], the waiting times are studied for the processor sharing service discipline, again under exponential assumptions. In Chap. 4 we extend this analysis to several other service disciplines.

In the work discussed above, the attention focuses on a fixed system. Another stream of work relates to systems in which the system size increases. An interesting observation to this end stems from Kleinrock. In [97], Sect. 4.11, he considers a repairman model with exponential working and service time distributions, with parameters μ and λ respectively. Kleinrock observes that for values of N that are much larger than μ/λ, the system behaves like a deterministic system, in which each additional machine causes all other machines to be delayed by its entire average service time.

This observation can be formalised through an approach based on stochastic process limits, see, e.g. Whitt [172]. In this approach, one considers a sequence of models with repair speed μ^N parameterised by the number of machines N. One then rescales the relevant processes in these models, such as the number of machines in repair, and studies these for N tending to infinity under the Kleinrock-type condition that N is larger than μ^N/λ. The limits of these rescaled processes then serve as rough approximations to these processes, and enable the derivation of some asymptotic characteristics that are not amenable to direct analysis. These approximations are then useful for large systems.

A pioneering study to this end is given in Iglehart [81]. He considers a repairman model with spare machines, a number of servers, and exponential service and working times. In this setting, he obtains a limit theorem that can be used to approximate the fluctuations in the number of machines in repair by means of an Ornstein-Uhlenbeck process. This result can then in turn be used to calculate transient characteristics. Results to this end can also be found in Mandelbaum and Pats [118] and Krichagina and Puhalskii [103]. The latter consider a closed queueing system which is equivalent to the repairman model without spares and one server. The speed of the server may depend on the number of machines in repair. Under the assumption of generally distributed working times and repair times that have finite mean and variance, they establish the desired approximations to the queue size processes. Our approach in Chap. 5 falls within this body of work on stochastic process limits.

In this book, we propose the repairman model as an appropriate model for the access delay due to contention trees in a finite population setting. This causes us to consider the repairman model with some new service disciplines: Random Order of Service, Gated Random Order of Service, and Gated Partial Random Order of Service, see Chap. 4. In Chap. 5, we build upon the asymptotic analysis in [103], to give an approximation to the sojourn time distribution in the repairman model with Gated Random Order of Service.

2.3 The Bulk Service Queue

Bulk service is illustrated in Fig. 2.4. The sow can feed up to a certain number of piggies simultaneously at a given time. The service discipline in a bulk service queue combines this notion of bulk service with a notion of periodicity. It is a discrete-time queueing model in which a central server periodically serves a fixed number of customers that arrive according to some random process.

The bulk service queue is defined via the recursion

$$X_{n+1} = (X_n - s)^+ + A_n. \tag{2.1}$$

Here, X_n denotes the queue size at the end of time slot n, A_n denotes the number of new arrivals during time slot n, the A_n are i.i.d. for all n, $\mathbb{E}(A_n) < s$, and $x^+ := \max(x, 0)$. The integer-valued quantity s denotes the fixed capacity of the server, i.e. the maximum size of the bulk served during one time slot.

In this book, the bulk service queue is used to represent the data queue of the data-transmission phase of a reservation procedure. It is defined as the (conceptual) queue that contains the packets for which transmission has been successfully requested but that have not yet been transmitted. In Chap. 6, we give a historical account of the developments concerning the bulk service queue. We discuss both its usage and the methodology to solve for the stationary distribution of the queue size.

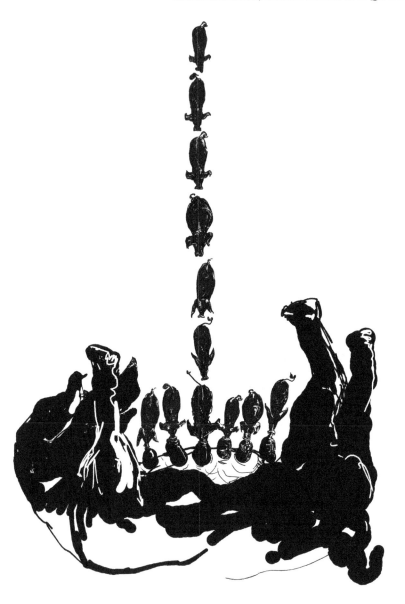

Fig. 2.4. Bulk service

One of our primary contributions to the bulk service queue is its adaptation for usage in a reservation procedure. For this, as an initial step, we write (2.1) as

$$X_{n+1} = (X_n - (f - c))^+ + \sum_{i=1}^{c} Y_{ni}, \qquad (2.2)$$

where the Y_{ni} form a sequence of i.i.d. random variables such that $c\mathbb{E}(Y_{ni}) < f - c$. With the obvious identifications $s = f - c$ and $A_n = \sum_{i=1}^{c} Y_{ni}$, recursion (2.2) reduces to (2.1).

Although (2.2) is just a trivial rewriting of (2.1), it is more suited to clarify the relevance of the bulk service queue for the data-transmission phase of a reservation procedure in cable networks, see Sect. 1.3. As explained there, such a procedure is frame-based, and each frame consists of f time slots. A number of these slots are devoted to the reservation procedure and a number of these slots are devoted to the actual transmission of messages for which a successful reservation has been made. Thus, in (2.2), it has been assumed that there are c slots for the reservation procedure of which the outcome adds $\sum_{i=1}^{c} Y_{ni}$ data packets to the data queue. The remaining slots can be used for data transmission, so that during each frame, up to $f - c$ packets are subtracted from the data queue. Clearly, X_n denotes the size of the data queue at the beginning of frame n. Figure 2.5 gives a graphical illustration of the process. The model defined by (2.2) is referred to as the *fixed boundary* model in the remainder of this monograph.

Clearly, if the data queue is empty at the beginning of a data slot, capacity is lost in the fixed boundary model. Therefore, the second model considered is one that designates the unused data slots as request slots, and is referred to as the *flexible boundary* model, which reflects the fact that the division of a frame into request and data slots can vary from one frame to another. This leads one to consider the recursion

$$X_{n+1} = (X_n - s)^+ + \sum_{i=1}^{c+(s-X_n)^+} Y_{ni}. \tag{2.3}$$

We refer to the c request slots that are scheduled at the beginning of every frame as *forced* request slots, and to the $(s - X_n)^+$ slots as *additional* request

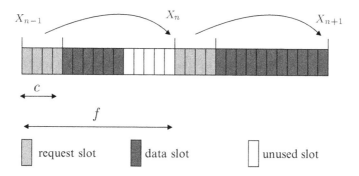

Fig. 2.5. The fixed boundary model. A frame of f slots consists of c request slots, followed by a maximum of $f - c$ data slots. Packets that arrive during frame n cannot depart from the queue until the beginning of frame $n + 1$

slots. Intuitively, the data-transmission schedule associated with the flexible boundary model is more efficient than the schedule that goes with the fixed boundary model, but one wants to have a clear quantitative understanding of its potential benefits. We will provide such understanding in Chap. 7 by analysing the packet delay in either model.

Recursion (2.3) still ignores the round-trip time that may be present in the communication channel. To include this delay, we introduce a delay parameter d and consider the following recursion:

$$X_{n+1} = (X_n - s)^+ + \sum_{i=1}^{c+(s-X_{n-d})^+} Y_{n-d,i}. \tag{2.4}$$

Finding the stationary distribution of the multi-dimensional Markov chain (2.4) is much harder than finding the stationary distributions of the one-dimensional Markov chains (2.2) and (2.3), see the appendix to Chap. 8.

For the fixed boundary model (2.2) we show in Chap. 7 that the probability generating function of the stationary queue length follows from the solution of the classical discrete bulk service queue. We next derive, using a more advanced technique, the probability generating function of the packet delay. From these transform solutions, the entire probability distributions can be obtained, as well as explicit expressions for performance characteristics like the mean and variance. For the flexible boundary model (2.3) we obtain similar results, although the derivation gets slightly more complicated. For both models we investigate the impact of the forced arrival slots c, in relation with other settings like the frame length and type of arrival process. In Chap. 8 we derive for the delayed bulk service queue (2.4) bounds and approximations to investigate the influence of c and d on the mean and variance of the stationary queue length. An exact analysis is reported in Appendix 8.B.

As a final model, we state a further generalisation of (2.4):

$$X_{n+1} = (X_n - (f - c_n))^+ + \sum_{i=1}^{c_n-d} Y_{n-d,i}. \tag{2.5}$$

In (2.5), c_n stands for the number of time slots devoted to the request process in frame n and $\{c_n, n \in \mathbb{N}\}$ is a sequence of integers, subject to the constraints $0 \le c_n \le f$. The model requires a scheduling policy to decide on the value of c_n. One such policy is implicit in (2.4), and uses

$$c_n = c + (f - c - X_n)^+. \tag{2.6}$$

In Chap. 8 we provide an alternative policy in which c_n is computed from the observed system values $c_{n-1}, c_{n-2}, \ldots$ and $X_n, X_{n-1}, \ldots$ and we show that it improves upon the policy in (2.6) in that it considerably reduces the expected data-queue size.

2.4 Tandem Queues with Shared Service Capacity

In Fig. 2.6, we illustrate 'shared service capacity', in which one server must divide its service capacity among various tasks. The tandem queues considered by us are two node networks. Users arrive at the first node, where they are served, and then move on to a second node. Upon service completion at the second node, they leave the network. In tandem queues with shared service capacity, the two nodes share a common capacity. Hence, these tandem queues can be used as a model for a reservation procedure, see Fig. 2.7, under the

Fig. 2.6. Shared service capacity

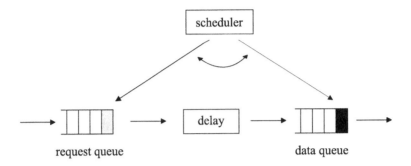

Fig. 2.7. Schematic view of the upstream channel of a cable network regulated by a reservation procedure

assumption that the delay is negligible. The capacity must be divided and we assume that a proportion p of the capacity is given to the first node, and $1 - p$ to the second node whenever both nodes are occupied.

The shared aspect shows up when one of the nodes is empty. First, we consider the case that the proportion of the capacity allocated to the first node is increased from p to 1, whenever the second node is empty. We will refer to this scheduling discipline as *partial coupling*. Note that this is an appropriate approximation to the flexible boundary model considered in Sect. 2.3 above: If there are no jobs in the data queue (second node), the number of time slots for the request queue (first node) is increased.

Under partial coupling, the service capacity of the first node depends on the amount of work at the second node, and this interdependence between the queues severely complicates the analysis. A natural extension of partial coupling is then full coupling, where not only the capacity of the first node is increased from p to 1 when the second node is empty, but the capacity of the second node is also increased from $1 - p$ to 1 when the first node is empty. Both partial and full coupling guarantee a minimum rate p for the first node and $1 - p$ for the second node, whenever there is work to be done at the node in question. However, contrary to partial coupling, full coupling is work-conserving in the sense that the service capacity is always fully used, irrespective of one of the queues being empty or not.

A service discipline that changes the service rates whenever one of the queues is empty is known in the queueing literature as *coupled processors*. If the coupled processors discipline is work-conserving, it reduces to full coupling. Full coupling is better known as *generalised processor sharing* (GPS). GPS is a popular scheduling discipline in modern communication networks, since it provides a way to achieve service differentiation among different types of traffic classes. For an overview of the literature on GPS we refer to Borst et al. [19], and the references therein. Throughout this book, we will refer to GPS/full coupling as coupled processors.

When we assume that users arrive to the first node according to a Poisson process, and that they require exponential service times at both nodes, no coupling results in a tandem queue of two independent $M/M/1$ queues. Since this is a standard Jackson network, the stationary joint queue length distribution possesses a pleasant product form, see p. 215 of this book.

This does not hold for partial and full coupling. These service disciplines give rise to two-dimensional Markov processes that can be solved using the theory of *boundary value problems*. This is because the joint queue length can be modelled as a random walk on the lattice in the first quadrant, and belongs as such to the class of nearest-neighbour random walks, in which only transitions to immediate neighbours may occur. A pioneering study of these types of random walks is the one of Malyshev [115], whose technique was introduced to queueing theory by Fayolle and Iasnogorodski [61]. They analysed two parallel queues with coupled processors, each queue having Poisson arrivals and exponential service times. They showed that the functional equation for the probability generating function of the joint queue length distribution can be transformed to a Riemann–Hilbert boundary value problem. Cohen and Boxma [39] have presented a systematic and detailed study of the technique of reducing a two-dimensional functional equation of a random walk or queueing model to a boundary value problem, and discuss in detail the numerical issues involved. In particular, the analytic solution to the boundary value problem requires the determination of some conformal mapping, which can be accomplished via the solution of singular integral equations. In most cases, this requires a numerical approach, see Cohen and Boxma [39].

Blanc [18] has investigated the transient behaviour of the ordinary two-station tandem queue, so without coupled processors. In his analysis, Blanc transforms the functional equation for the probability generating function of the joint queue length distribution into a Riemann–Hilbert boundary value problem, using the same technique as introduced by Fayolle and Iasnogorodski [61]. For the two-stage tandem queue with coupled processors, Resing and Örmeci [148] made a similar transformation. Other applications of the theory of boundary value problems to queueing models can be found in [16, 33, 38, 39, 62, 63, 96, 123, 130] and the references therein.

For the *two-stage tandem queue with coupled processors* we show in Chap. 9 that the problem of finding the generating function of the joint stationary queue length distribution can be reduced to two different Riemann–Hilbert boundary value problems. We discuss the similarities and differences between the two boundary value problems, and relate them to the computational aspects of obtaining performance measures like the mean queue length and the fraction of time a queue is empty. Our detailed account of the numerical issues that arise when implementing a formal solution to a Riemann–Hilbert boundary value problem, is illustrative and may serve as an example for other types of queues that can be solved using the same technique. For the *two-stage tandem queue with partial coupling* we will show that the problem of finding the bivariate generating function of the joint stationary queue length distribution can be reduced to a Riemann–Hilbert boundary value problem of a slightly different type. The solution to this boundary value problem is more involved than the one for the coupled processors discipline. We indicate how the solution to the model with partial coupling can be obtained, but we do not discuss all the details.

Next, in Chap. 10, we present a more general model of a *two-station network with coupled processors*. After receiving service at a server, a user either joins the queue of the same server, joins the queue of the other server, or leaves the system, each with a given probability. Users require exponential service times at each station. This general network model covers both the model of Fayolle and Iasnogorodski [61] and the two-stage tandem queue with coupled processors as special cases. We show that the general model can be solved using the theory of boundary value problems. We also consider the case that one of the queues has preemptive priority over the other queue. For this priority case, we show that the generating function of the joint stationary queue length distribution can be obtained directly from the functional equation without employing the theory of boundary value problems.

PART II

CONTENTION TREES

Chapter 3

BASIC PROPERTIES OF CONTENTION TREES

Contention trees constitute a popular class of algorithms to provide access to a shared resource for a random population of contenders. In particular, they can be used to transmit the request messages of a reservation procedure. In this chapter, we explore the basic properties of contention trees. The main original contribution of this chapter concerns our study of the delay experienced in the use of a contention tree. For this we introduce the notion of success instant. We give an approximation to the marginal empirical distribution of a success instant and show that this approximation is asymptotically exact for an increasing number of contenders.

3.1   Introduction

The use of contention trees was reviewed in Sect. 2.1. In this chapter, we will ignore the details of the way in which contention trees are used, and concentrate on the basic properties that are common to all applications. Thus, we confine attention to the basic m-ary tree to resolve a fixed number, n, of conflicts.

In such a tree, a group of colliding messages is recursively split into subgroups until all collisions are resolved. This recursive splitting can be depicted graphically by means of a tree, and an example tree is given in Fig. 3.1 for $m = 3$ and $n = 8$. The tree represents a collection of slots of the communication channel plus the number of transmission attempts in these slots. If the number associated with a slot equals 0, then there is no transmission attempt in this slot. The value 1 indicates that there is exactly one transmission in that slot, and in this case we have a successful transmission. A value greater than 1 indicates multiple transmission attempts. Consequently, all messages are lost and must be retransmitted. This is done in the following way. A group of m slots is designated as the child node of the slot in which the collision occurred. Any colliding message is retransmitted in one of the slots in the child node, where

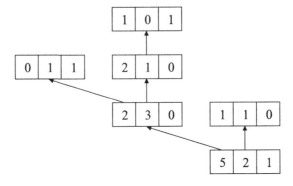

Fig. 3.1. Basic contention tree: Slots of the tree with a collision, i.e. an entry $\rangle 1$, are recursively split until all slots are empty (0) or have a successful transmission (1)

5	2	1	2	3	0	1	1	0	0	1	1	2	1	0	1	0	1

Fig. 3.2. Same tree as in Fig. 3.1, with a breadth-first ordering of the slots

5	2	0	1	1	3	2	1	0	1	1	0	0	2	1	1	0	1

Fig. 3.3. Same tree as in Fig. 3.1, with a depth-first ordering of the slots

the slot is chosen at random and independently of the choice of the other messages. This splitting continues until all slots are either empty or contain one, successful, transmission.

Contention trees have been well studied and we refer to Sect. 2.1 for an overview of the literature. The statistic that has received the most attention is the average length of the tree, given that initially n contenders collide in the tree. In Janssen and de Jong [86], two additional statistics are introduced: The number of levels required for a random contender to be successful and the number of levels required to complete the tree. For all statistics, expressions are given for the first two moments.

In this chapter, we study yet another statistic of the contention tree: The success instant of a random station. To motivate this statistic, observe that, in practice, the nodes correspond to the time slots of the communication channel devoted to the contention resolution. So, the slots of the tree must be time-ordered. Figures 3.2 and 3.3 represent the same tree as shown in Fig. 3.1, but now time-ordered. Figure 3.2 displays a tree ordered according to the breadth-first ordering, in which the nodes are ordered according to their distance to the root node, see Sect. 3.2.2 for a formal definition. Figure 3.3 illustrates the depth-first (or stack) ordering, in which the subtree starting with the first slot of a node occurs before the subtree starting with the second slot of that node.

Using either time ordering, we can speak of success instants: The instants of the successful transmissions relative to the start of the tree. In Fig. 3.2, the success instants occur in slots 3, 7, 8, 11, 12, 14, 16, and 18. In Fig. 3.3, the success instants occur in slots 4, 5, 8, 10, 11, 15, 16, and 18.

The main contribution in this chapter concerns our study of the delay experienced in using a contention tree, and in particular the expression for the marginal distribution of the success instants. We confine ourselves to the case of trees traversed in breadth-first order, as this is the more efficient ordering in case of a communication channel characterised by large round-trip times, see Massey [119]. To prepare the ground we introduce in Sect. 3.2 two formal methods to describe contention trees. Moreover, we will derive the marginal and bivariate distributions, under the random splitting rule, of some of the variables associated with these models. Next, in Sect. 3.3.1, we will use these models to review some well known facts about contention trees. In Sect. 3.3.2, we turn to the success instants. We will give an approximation to the marginal distribution of the success instants, and show that this approximation is asymptotically exact for an increasing number of contenders. In Sect. 3.3.3, we consider the optimisation of contention trees via the introduction of skip levels. We give proofs of the various theorems in Sect. 3.4. We conclude with a brief summary, a generalisation to depth-first trees, and some ideas for further research.

3.2 Formal Tree Models

In this section, we present two formal ways to describe a contention tree. Both methods are based on the complete m-ary tree, also considered in Capetanakis [30] and Kaplan and Gulko [93]. The method of Sect. 3.2.1 associates a number, the subgroup size, with each slot in a complete tree. Next, in Sect. 3.2.2, we present a method that is based on infinite paths through a complete m-ary tree. With the random splitting rule introduced in the introduction, the variables associated with these descriptions become random variables. In Sect. 3.2.3, we provide expressions for their marginal and bivariate distributions.

3.2.1 A Slot Based Description

The most obvious way to describe a contention tree is by associating a number, the number of colliding messages, with each slot of the tree. To identify the slots, we use strings over the symbols $\{1, \ldots, m\}$. Let $\mathcal{J}_d$ denote the set of strings over $\{1, \ldots, m\}$ of length d:

$$\mathcal{J}_d := \{1, \ldots, m\}^d,$$

and let $\mathcal{J}$ denote the set of strings of arbitrary length:

$$\mathcal{J} := \bigcup_{d=0}^{\infty} \mathcal{J}_d.$$

Each string in $\mathcal{J}$ can be interpreted in the obvious way as the address of a slot in the complete tree, and the empty set, corresponding to $d = 0$, stands for the root slot of the tree. We use z_+ as the set of strings that is obtained from $z = \langle z_1 \ldots z_d \rangle \in \mathcal{J}$ by appending one symbol from $\{1, \ldots, m\}$ to z:

$$z_+ := \{\langle z_1 \ldots z_d\, 1\rangle, \ldots, \langle z_1 \ldots z_d\, m\rangle\},$$

and z_+ is interpreted as the set of child slots of slot z. Similarly, we define z_* as the set of strings that is obtained from z by appending any element $y \in \mathcal{J} \setminus \emptyset$, and z_* is interpreted as the set of descendants of z in the tree.

Using this notation, we define a tree description as a mapping $N : \mathcal{J} \to \mathbb{N}$ which satisfies the following condition for each $z \in \mathcal{J}$:

$$N(z) = \sum_{y \in z_+} N(y). \tag{3.1}$$

Thus, N associates a number with each string, subject to a conservation property. The number $N(z)$ represents the number of messages that collided in the slot associated with z. The conservation property states that each set of colliding messages is partitioned into subsets. Note here that (3.1) is a condition on N, as in Neininger and Rüschendorf [132], and not its definition.

REMARK 3.1 The function N codes all the information in a contention tree, so that it is a proper description. However, N is a redundant description as it is defined on all of $\mathcal{J}$. An actual contention tree differs from the complete m-ary tree in that the branching terminates as soon as a slot contains a value equal to 0 or 1. In the definition of N, the splitting continues forever. A minimal description of the contention tree, is given by the set of terminating slots plus the associated function values. Using these, we can reconstruct the value for each slot in the contention tree by summing over the values associated with its child slots, via (3.1).

These concepts are illustrated in Table 3.1 for the tree shown in Fig. 3.1. The left column of Table 3.1 lists a number of strings in $\mathcal{J}$ which stand for the slots in the tree. The right column lists the value of N for these strings. As noted in Remark 3.1, not all the slots of the complete m-ary tree are actually part of the contention tree. Referring to the tree shown in Fig. 3.1, we see that not all slots at level 2 are relevant to the contention tree: As $N(3) = 1$, the slots $\langle 31 \rangle$, $\langle 32 \rangle$, and $\langle 33 \rangle$ are not contained in the tree. Hence, Table 3.1 does not list the values

Table 3.1. All slots $z \in \mathcal{J}$ actually included in the contention tree displayed in Fig. 3.1, with value $N(z)$

Slot address z	Subgroup size $N(z)$
$\langle 1 \rangle$	5
$\langle 2 \rangle$	2
$\langle 3 \rangle$	1
$\langle 11 \rangle$	2
$\langle 12 \rangle$	3
$\langle 13 \rangle$	0
$\langle 21 \rangle$	1
$\langle 22 \rangle$	1
$\langle 23 \rangle$	0
$\langle 111 \rangle$	0
$\langle 112 \rangle$	1
$\langle 113 \rangle$	1
$\langle 121 \rangle$	2
$\langle 122 \rangle$	1
$\langle 123 \rangle$	0
$\langle 1221 \rangle$	1
$\langle 1222 \rangle$	0
$\langle 1223 \rangle$	1

of N for all strings in $\mathcal{J}$. Rather, it lists those that are relevant to the contention tree from Fig. 3.1. Note in particular that $N(\emptyset) = 8$ for this example. However, we have not included $\emptyset$ and $N(\emptyset)$ in Table 3.1, as we do not consider the root slot as part of the contention tree.

3.2.2 A Path Based Description

In Remark 3.1, we have observed that a contention tree can be reconstructed if we retain the locations of the terminal slots plus the values of N for these slots. In this section, we use this observation to give an alternative description of a contention tree, which is based on these stopping times on the paths through the complete tree.

We use $\mathcal{X}$ to denote the set of infinite strings over the alphabet $\{1, \ldots, m\}$:

$$\mathcal{X} = \{1, \ldots, m\}^{\infty},$$

and $\mathcal{X}$ codes the set of all infinite paths through the complete m-ary tree in the obvious way. For $x \in \mathcal{X}$, we use $x(d)$ to denote the restriction of x to its first d symbols, so that $x(d) \in \mathcal{J}_d$. By convention, $x(0) = \emptyset$.

The alternative description is specified via functions $\tau : \mathcal{X} \rightarrow \mathbb{N}$ and $\psi : \mathcal{X} \rightarrow \{0, 1\}$. Informally, $\tau(x)$ indicates the level at which the path x is stopped. Hence, if $\tau(x) = d$, then $x(d)$ provides the address of a slot in the contention

Table 3.2. All paths x through a complete ternary contention tree, where '*' is used as shorthand for all possible strings with symbols 1, 2, and 3, and the values of the functions τ and ψ for the contention tree shown in Fig. 3.1

	Paths	
x	$\tau(x)$	$\psi(x)$
$\langle 3*\rangle$	1	1
$\langle 13*\rangle$	2	0
$\langle 21*\rangle$	2	1
$\langle 22*\rangle$	2	1
$\langle 23*\rangle$	2	0
$\langle 111*\rangle$	3	0
$\langle 112*\rangle$	3	1
$\langle 113*\rangle$	3	1
$\langle 122*\rangle$	3	1
$\langle 123*\rangle$	3	0
$\langle 1221*\rangle$	4	1
$\langle 1222*\rangle$	4	0
$\langle 1223*\rangle$	4	1

tree that contains either a successful transmission or that is empty. The success indicator $\psi(x)$ is 0-1 valued and equals 0 if the path was stopped by an empty slot and equals 1 if the path was stopped by a successful transmission.

Formally, τ and ψ are obtained from N: For any $x \in \mathcal{X}$ define

$$\tau(x) := \inf\left(d \in \mathbb{N} : N(x(d-1)) > 1, N(x(d)) \le 1\right), \qquad (3.2)$$

and

$$\psi(x) := N(x(\tau(x))). \qquad (3.3)$$

Table 3.2 illustrates these concepts using the tree shown in Fig. 3.1. In Table 3.2, we list all strings in $\mathcal{X}$. Here, we have shortened this infinite list by using the symbol '*' for all possible strings of the symbols 1, 2, and 3. Next to these paths we have listed the values of τ and ψ.

REMARK 3.2 The descriptions via N and (τ, ψ) are equivalent in the sense that they code the same contention tree, as is evident from Remark 3.1. Using N, we make some arbitrary (and unimportant) choice of what happens in slots of the complete tree that are descendants of slots with successful transmissions, using τ and ψ such a choice is not needed.

In dealing with trees, we make extensive use of an ancestry relation, $\sim$, between elements of $\mathcal{J}$. Given $u \in \mathcal{J}$ and $v \in \mathcal{J}$, we use $u \sim v$ to denote the situation that either $u = v$, $u \in v_*$, or $v \in u_*$. The case that neither of these conditions holds, is indicated as $u \nsim v$. In particular, we need to count the fraction of paths in $\mathcal{X}$ that satisfy some specific ancestry relation. The following lemma gives the relevant facts, where we use U to denote the uniform measure on $\mathcal{X}$, and $\mathbf{1}$ for the indicator function.

LEMMA 3.1 *For* $1 \leq d \leq e$ *we have*

$$\sum_{x \in \mathcal{X}} \sum_{y \in \mathcal{X}} \mathbf{1}(x(d) \sim y(e)) U(x) U(y) = \frac{m^e}{m^{d+e}}$$

and

$$\sum_{x \in \mathcal{X}} \sum_{y \in \mathcal{X}} \mathbf{1}(x(d) \nsim y(e); x(d-1) \sim y(e-1)) U(x) U(y) = \frac{m^{e+1} - m^e}{m^{d+e}}.$$

The proof of Lemma 3.1 is a simple counting argument, and not further considered.

3.2.3 Distributions Under the Random Splitting Rule

Hitherto, we have used the models as descriptions, i.e. as a means to code a given contention tree. However, with the splitting rule as indicated in the introduction, a contention tree is a stochastic object. In particular, the description $\{N(z), z \in \mathcal{J}\}$ is a collection of random variables. For an arbitrary integer n, its distribution can be defined as follows:

$$N(\emptyset) = n,$$
$$\{N(y), y \in z_+\} \stackrel{d}{=} \text{Mult}(N(z), 1/m, \dots, 1/m), \tag{3.4}$$

where $\stackrel{d}{=}$ denotes equality in distribution and Mult denotes the multinomial distribution with all probabilities equal to $1/m$.

We will now use this recursive definition to give a non-recursive definition of the marginal and bivariate distributions of the elements of $\{N(z), z \in \mathcal{J}\}$. This will be the first goal of the section, see Lemma 3.2. The distribution of $\{N(z), z \in \mathcal{J}\}$ induces the distribution of $\{(\tau(x), \psi(x)), x \in \mathcal{X}\}$ via the definitions (3.2) and (3.3). We survey what we need of their joint distribution in Lemmas 3.3 and 3.4. Proofs of the lemmas are given at the end of this chapter. This section ends with a remark on a scaling behaviour of these distributions that provides a rationale for the further developments.

First, we observe that the definition (3.4) leads to analytic expressions for the marginal distribution of each $N(z)$ and for the bivariate distributions of the pairs $N(u)$ and $N(v)$:

LEMMA 3.2 *(a) If* $d \geq 1$ *and* $z \in \mathcal{J}_d$, *then for* $0 \leq i \leq n$,

$$\mathbb{P}(N(z) = i) = \binom{n}{i} \left(\frac{1}{m^d} \right)^i \left(1 - \frac{1}{m^d} \right)^{n-i}. \tag{3.5}$$

(b) If $1 \leq d \leq e$, $u \in \mathcal{J}_d$ and $v \in \mathcal{J}_e$ with $u \sim v$, then for $0 \leq i \leq j \leq n$,

$$\mathbb{P}(N(u) = j, N(v) = i) \tag{3.6}$$

$$= \binom{n}{i, j-i, n-j} \left(\frac{1}{m^e}\right)^i \left(\frac{1}{m^d} - \frac{1}{m^e}\right)^{j-i} \left(1 - \frac{1}{m^d}\right)^{n-j}.$$

(c) If $1 \leq d \leq e$, $u \in \mathcal{J}_d$ and $v \in \mathcal{J}_e$ with $u \not\sim v$, then for $0 \leq i \leq n$ and $0 \leq j \leq n$ with $i + j \leq n$,

$$\mathbb{P}(N(u) = i, N(v) = j) \tag{3.7}$$

$$= \binom{n}{i, j, n-i-j} \left(\frac{1}{m^d}\right)^i \left(\frac{1}{m^e}\right)^j \left(1 - \frac{1}{m^d} - \frac{1}{m^e}\right)^{n-i-j}.$$

The marginal distribution of $N(z)$ stated in Lemma 3.2(a) can be derived via a direct probabilistic argument: The n contenders are distributed uniformly and independently of each other over the m^d slots at level d. Similar arguments can be brought to bear on the bivariate distributions given in Lemma 3.2. The proof at the end of this chapter, however, uses formal arguments based on induction.

Now from these basic probabilities we can deduce the marginal and bivariate distributions related to τ. To this end, define for $d \geq 1$,

$$q(d) := 1 - \left(1 - \frac{1}{m^{d-1}}\right)^n - \frac{n}{m^{d-1}} \left(1 - \frac{1}{m^{d-1}}\right)^{n-1}, \tag{3.8}$$

and for $d \geq 1$ and $e \geq 1$,

$$\begin{aligned} n(d, e) \; := \; & 1 - \left(1 - \frac{1}{m^{d-1}}\right)^n - \frac{n}{m^{d-1}} \left(1 - \frac{1}{m^{d-1}}\right)^{n-1} \\ & - \left(1 - \frac{1}{m^{e-1}}\right)^n - \frac{n}{m^{e-1}} \left(1 - \frac{1}{m^{e-1}}\right)^{n-1} \\ & + \left(1 - \frac{1}{m^{d-1}} - \frac{1}{m^{e-1}}\right)^n \\ & + n\left(\frac{1}{m^{d-1}} + \frac{1}{m^{e-1}}\right) \left(1 - \frac{1}{m^{d-1}} - \frac{1}{m^{e-1}}\right)^{n-1} \\ & + n(n-1)\frac{1}{m^{d-1}} \frac{1}{m^{e-1}} \left(1 - \frac{1}{m^{d-1}} - \frac{1}{m^{e-1}}\right)^{n-2}. \end{aligned} \tag{3.9}$$

LEMMA 3.3 *(a) If $d \geq 1$ and $x \in \mathcal{X}$,*

$$\mathbb{P}(\tau(x) \geq d) = q(d). \tag{3.10}$$

(b) If $1 \leq d \leq e$, and $x, y \in \mathcal{X}$ with $x(d-1) \sim y(e-1)$,

$$\mathbb{P}(\tau(x) \geq d, \tau(y) \geq e) = q(e). \tag{3.11}$$

(c) *If $d \geq 1$ and $e \geq 1$, and $x, y \in \mathcal{X}$ with $x(d-1) \not\sim y(e-1)$,*

$$\mathbb{P}(\tau(x) \geq d, \tau(y) \geq e) = n(d, e). \tag{3.12}$$

Again, as in Lemma 3.2, some of the distributions in Lemma 3.3 can be obtained via a direct probabilistic argument. The distribution in (3.10) immediately follows from the observation that a given slot at level d is in the actual contention tree if the subgroup size associated with its predecessor, at level $d - 1$, is at least 2. To establish (3.11), combine the previous argument with the observation that in this case $\tau(y) \geq e$ implies $\tau(x) \geq d$. The distribution given in (3.12) requires a somewhat more elaborate argument, see [93]. At the end of this chapter, we give proofs using induction.

We can also consider the marginal and bivariate distributions of the success instants along a given path. To this end define $I_1(x, d)$ as the event that path $x \in \mathcal{X}$ has a successful exit at level d:

$$I_1(x, d) := \{\tau(x) = d, \psi(x) = 1\}. \tag{3.13}$$

Moreover, define for $d \geq 1$,

$$q_1(d) \ := \ \frac{n}{m^d}\left(\left(1 - \frac{1}{m^d}\right)^{n-1} - \left(1 - \frac{1}{m^{d-1}}\right)^{n-1}\right), \tag{3.14}$$

$$c_0(d) \ := \ \frac{n(n-1)}{m^{2d}}\left(1 - \frac{2}{m^d}\right)^{n-2}, \tag{3.15}$$

for $1 \leq d < e$,

$$c_1(d, e) := \frac{n(n-1)}{m^{d+e}}\left(\left(1 - \frac{1}{m^d} - \frac{1}{m^e}\right)^{n-2} - \left(1 - \frac{1}{m^d} - \frac{1}{m^{e-1}}\right)^{n-2}\right),$$

and for $d \geq 1$ and $e \geq 1$,

$$n_1(d, e) \ := \ \frac{n(n-1)}{m^{e+d}}\left(\left(1 - \frac{1}{m^d} - \frac{1}{m^e}\right)^{n-2} - \left(1 - \frac{1}{m^{d-1}} - \frac{1}{m^e}\right)^{n-2}\right.$$

$$\left. - \left(1 - \frac{1}{m^d} - \frac{1}{m^{e-1}}\right)^{n-2} + \left(1 - \frac{1}{m^{d-1}} - \frac{1}{m^{e-1}}\right)^{n-2}\right).$$

LEMMA 3.4 (a) *If $d \geq 1$ and $x \in \mathcal{X}$,*

$$\mathbb{P}(\tau(x) = d, \psi(x) = 1) = q_1(d). \tag{3.16}$$

(b) *If $1 \leq d \leq e$, and $x, y \in \mathcal{X}$ with $x(d) \sim y(e)$,*

$$\mathbb{P}(I_1(x, d), I_1(y, e)) = q_1(d)\mathbf{1}(d = e). \tag{3.17}$$

(c) If $d \geq 1$, and $x, y \in \mathcal{X}$ with $x(d) \not\sim y(d)$ and $x(d-1) \sim y(d-1)$,

$$\mathbb{P}(I_1(x, d), I_1(y, d)) = c_0(d). \tag{3.18}$$

(d) If $1 \leq d < e$, and $x, y \in \mathcal{X}$ with $x(d) \not\sim y(e)$ and $x(d-1) \sim y(e-1)$,

$$\mathbb{P}(I_1(x, d), I_1(y, e)) = c_1(d, e). \tag{3.19}$$

(e) If $1 \leq d$, $1 \leq e$, and $x, y \in \mathcal{X}$ with $x(d-1) \not\sim y(e-1)$,

$$\mathbb{P}(I_1(x, d), I_1(y, e)) = n_1(d, e). \tag{3.20}$$

REMARK 3.3 Recall that for $n \to \infty$

$$\lim_{n \to \infty} \left(1 - \frac{x}{n}\right)^n = e^{-x}, \tag{3.21}$$

and consequently,

$$\lim_{k \to \infty} \left(1 - \frac{1}{m^{k+\delta}}\right)^{\xi m^k} = e^{-\xi m^{-\delta}}. \tag{3.22}$$

Hence, with $n = \xi m^k, \xi \in [1, m)$ and $\delta \geq -k + 1$, we find for q as defined in (3.8) that

$$\lim_{k \to \infty} q(k + \delta + 1) = 1 - e^{-\xi m^{-\delta}} - \xi m^{-\delta} e^{-\xi m^{-\delta}}. \tag{3.23}$$

Now consider

$$\mathbb{P}(\tau \geq k + \delta + 1) = q(k + \delta + 1) \tag{3.24}$$

for $n = \xi m^k$ and $k \to \infty$. From (3.23) it follows that, asymptotically, the probability in (3.24) does not depend on k. Inspection of the bivariate probabilities in Lemmas 3.3 and 3.4 reveals that these bivariate probabilities share this asymptotic independence of k.

We can interpret this scale invariance heuristically: Increasing n by increasing k shifts the level at which events occur. However, it does not change the univariate and bivariate probabilities of these events in the sense that the event $I_1(\langle v_1 \ldots v_d \rangle)$ for $n = \xi m^k$ follows the same probability distribution as the event $I_1(\langle v_1 \ldots v_d v_{d+1} \ldots v_{d+l} \rangle)$ for $n = \xi m^{k+l}$. However, there are many more events of the latter type. Indeed, the contention tree with $n = m^{k+l}$ consists of m^l probabilistic copies of the contention tree with $n = m^k$ shifted downward by l levels. Therefore, one can expect a strong law of large numbers to hold at levels $k + l$ for $l \to \infty$. This underlies much of the theory in Sects. 3.3.1 and 3.3.2.

3.3 Tree Statistics

We give a brief account of the most extensively studied statistic of the tree: Its length. Then we turn to the success instants in Sect. 3.3.2. Finally, in Sect. 3.3.3, we consider tree statistics for trees in which a number of the initial levels has been skipped.

3.3.1 Tree Length

The tree length equals the number of slots that are needed to resolve all conflicts in the tree. A simple calculation shows that the tree length T can be expressed as

$$T = \sum_{l=1}^{\infty} m^l \sum_{x \in \mathcal{X}} \mathbf{1}(\tau(x) \geq l) U(x). \tag{3.25}$$

To verify (3.25), observe that a node at level l is shared by a fraction m^{-l} of all the paths in $\mathcal{X}$.

Taking expectations, we obtain an expression for the expected tree length:

$$\mathbb{E}(T) = \mathbb{E}\left(\sum_{l=1}^{\infty} m^l \sum_{x \in \mathcal{X}} \mathbf{1}(\tau(x) \geq l) U(x) \right) = \sum_{l=1}^{\infty} m^l q(l), \tag{3.26}$$

where we have used (3.10), also see Kaplan and Gulko [93], (1). Now evaluating $\mathbb{E}(T)/n$ for $n = \xi m^k$, as motivated by Remark 3.3, we obtain

$$\begin{aligned}
\frac{1}{n}\mathbb{E}(T) &= \frac{1}{\xi m^k} \sum_{l=1}^{\infty} m^l q(l) \\
&= m \sum_{l=1}^{\infty} \frac{1}{\xi} m^{-k+l-1} q(k + (-k + l - 1) + 1) \\
&= m \sum_{i=-k}^{\infty} \frac{1}{\xi} m^i q(k + i + 1).
\end{aligned}$$

Using the asymptotic form of $q(k + i + 1)$ given in (3.23), it is not difficult to obtain the following expression for the expected tree length.

THEOREM 3.1 (Adapted from [93], Theorem 1). *For* $n = \xi m^k$, *with* $\xi \in [1, m)$,

$$\lim_{k \to \infty} \frac{1}{n}\mathbb{E}(T) = m a(\xi), \tag{3.27}$$

where

$$a(\xi) = \sum_{i=-\infty}^{\infty} h\left(\frac{\xi}{m^i}\right), \quad and \quad h(u) = \frac{1 - e^{-u}}{u} - e^{-u}. \tag{3.28}$$

Expression (3.27) provides a convenient starting point to analyse the properties of the expected tree length. In [86] it is shown that

$$\mathbb{E}(T) \approx m \left(\frac{n}{\log(m)} - \frac{1}{m-1} \right) \tag{3.29}$$

is a good approximation to $\mathbb{E}(T)$ for n large. However, numerical computations readily reveal that $a(\xi)$ is not constant over the range $[1, m]$. From this we deduce the remarkable fact, see e.g. [86, 93, 163], that $\mathbb{E}(T)/n$ does not converge to a fixed value in the limit with n tending to ∞. Rather $\mathbb{E}(T)/n$ exhibits a tiny oscillation around the leading term given in (3.29). In [86], this oscillation is identified.

We now turn to the variance of the tree length. For this consider T^2:

$$
\begin{aligned}
T^2 &= \left(\sum_{d=1}^{\infty} m^d \sum_{x \in \mathcal{X}} \mathbf{1}(\tau(x) \geq d)\mathrm{U}(x) \right)^2 \\[2mm]
&= \sum_{d=1}^{\infty} m^{2d} \sum_{x \in \mathcal{X}} \sum_{y \in \mathcal{X}} \mathbf{1}(\tau(x) \geq d, \tau(y) \geq d)\mathrm{U}(x)\mathrm{U}(y) \\[2mm]
&\quad + 2 \sum_{d=1}^{\infty} \sum_{e=d+1}^{\infty} m^d m^e \sum_{x \in \mathcal{X}} \sum_{y \in \mathcal{X}} \mathbf{1}(\tau(x) \geq d, \tau(y) \geq e)\mathrm{U}(x)\mathrm{U}(y).
\end{aligned}
\tag{3.30}
$$

Splitting the sums in (3.30) gives

$$
\begin{aligned}
T^2 &= \sum_{d=1}^{\infty} m^{2d} \sum_{\substack{x,y \in \mathcal{X} \\ x(d-1) \sim y(d-1)}} \mathbf{1}(\tau(x) \geq d, \tau(y) \geq d)\mathrm{U}(x,y) \\[2mm]
&\quad + \sum_{d=1}^{\infty} m^{2d} \sum_{\substack{x,y \in \mathcal{X} \\ x(d-1) \not\sim y(d-1)}} \mathbf{1}(\tau(x) \geq d, \tau(y) \geq d)\mathrm{U}(x,y) \\[2mm]
&\quad + 2 \sum_{d=1}^{\infty} \sum_{e=d+1}^{\infty} m^{d+e} \sum_{\substack{x,y \in \mathcal{X} \\ x(d-1) \sim y(e-1)}} \mathbf{1}(\tau(x) \geq d, \tau(y) \geq e)\mathrm{U}(x,y) \\[2mm]
&\quad + 2 \sum_{d=1}^{\infty} \sum_{e=d+1}^{\infty} m^{d+e} \sum_{\substack{x,y \in \mathcal{X} \\ x(d-1) \not\sim y(e-1)}} \mathbf{1}(\tau(x) \geq d, \tau(y) \geq e)\mathrm{U}(x,y).
\end{aligned}
$$

Here, we have used $\mathrm{U}(x, y)$ as shorthand for $\mathrm{U}(x)\mathrm{U}(y)$. The expectations of the indicator variables are given in Lemma 3.3. Moreover, the fraction of paths such that $x(d-1){\sim}y(e-1)$ or $x(d-1){\not\sim}y(e-1)$ is given in Lemma 3.1. Consequently, the expected value of (3.30) equals

$$
\mathbb{E}(T^2) = \sum_{d=1}^{\infty} \left[q(d)m^{d+1} + n(d,d)\left(m^{2d} - m^{d+1}\right) \right]
$$
$$
+2 \sum_{d=1}^{\infty} \sum_{e=d+1}^{\infty} \left[q(e)m^{e+1} + n(d,e)(m^{d+e} - m^{e+1}) \right].
$$

An expression for $\mathrm{var}(T)$ can be obtained by subtracting $(\mathbb{E}T)^2$ and splitting the terms of $\mathbb{E}(T^2)$ as above. So we obtain

$$
\begin{aligned}
\mathrm{var}(T) &= \mathbb{E}(T^2) - (\mathbb{E}T)^2 \qquad\qquad (3.31)\\
&= \sum_{d=1}^{\infty} \Big[\left(q(d) - q(d)^2 \right) m^{d+1} \\
&\qquad\quad + \left(n(d,d) - q(d)^2 \right) m^{d+1}(m^{d-1} - 1) \Big] \\
&\quad +2 \sum_{d=1}^{\infty} \sum_{e=d+1}^{\infty} [q(e) - q(d)q(e)]\, m^{e+1} \\
&\quad +2 \sum_{d=1}^{\infty} \sum_{e=d+1}^{\infty} [n(d,e) - q(d)q(e)]\,(m^{d+e} - m^{e+1}).
\end{aligned}
$$

Now proceeding as above for $\mathbb{E}(T)$, we can obtain an asymptotic expression for $\mathrm{var}(T)/n$. Thus, on setting $n = \xi m^k$, letting k tend to ∞, and using the asymptotic form of the bivariate probabilities $n(d,e)$ and $q(d)$, it is not difficult to obtain the following theorem.

THEOREM 3.2 (See [93], Theorem 2). *For $n = \xi m^k$, with $\xi \in [1, m)$,*

$$
\lim_{k\to\infty} \frac{1}{n}\mathrm{var}(T) = m^2 b(\xi), \qquad\qquad (3.32)
$$

where

$$
\begin{aligned}
b(\xi) &= \sum_{i=-\infty}^{\infty} \left(1 + \xi m^{-i}\right) e^{-\xi m^{-i}} h\left(\xi m^{-i}\right) - \left(\sum_{i=-\infty}^{\infty} \xi m^{-i} e^{-\xi m^{-i}} \right)^2 \\
&\quad +2 \sum_{i=-\infty}^{\infty} \left(1 + \xi m^{-i}\right) e^{-\xi m^{-i}} \sum_{j>i} h\left(\xi m^{-j}\right)
\end{aligned}
$$

and h is as in Theorem 3.1.

As b is bounded and non-constant over the range $\xi \in [1, m)$, Theorem 3.2 shares some of the properties of Theorem 3.1. Firstly, we observe that $\text{var}(T)$ is of $O(n)$ for n large. Secondly, $\text{var}(T)/n$ does not converge to a fixed limit for n tending to ∞. Rather, $\text{var}(T)$ oscillates around its leading term. Again, the oscillations are tiny. In [86] explicit formulas are given, both for the leading term of $\text{var}(T)/n$ and for the oscillations around this main value.

We can also define various other tree statistics of interest. For example, by restricting the sum over l in (3.25) to the first d levels for the length of a tree, we obtain an expression for T_d, the number of slots in the contention tree up to level d:

$$T_d := \sum_{l=1}^{d} m^l \sum_{x \in \mathcal{X}} \mathbf{1}(\tau(x) \geq l) \mathrm{U}(x). \tag{3.33}$$

We can give similar expressions for the number of exits S from the tree:

$$S := \sum_{l=1}^{\infty} m^l \sum_{x \in \mathcal{X}} \mathbf{1}(\tau(x) = l, \psi(x) = 1) \mathrm{U}(x), \tag{3.34}$$

and for S_d, the number of exits from the contention tree up to level d:

$$S_d := \sum_{l=1}^{d} m^l \sum_{x \in \mathcal{X}} \mathbf{1}(\tau(x) = l, \psi(x) = 1) \mathrm{U}(x). \tag{3.35}$$

Clearly, we must have that $S = n$. Taking expectations in (3.34) and substituting the expression for $q_1(d)$, we find that $\mathbb{E}(S) = n$. Moreover, a lengthy but otherwise elementary computation, involving the same techniques that lead to the computation of $\text{var}(T)$, readily shows that $\text{var}(S) = 0$; see, e.g. the proof of Theorem 3.4.

3.3.2   Success Instants

Denote the success instants, as introduced in Sect. 3.1, by $e_i, i = 1, \ldots, n$. Thus e_i is the time of the ith success in the contention tree, and this time will clearly depend on the order in which the tree is traversed. The empirical distribution of the exit times, $\hat{F}$ is given by

$$\hat{F}(t) = \frac{1}{n} \sum_{i=1}^{n} \mathbf{1}(e_i \leq t). \tag{3.36}$$

In this section, we will obtain approximations of $\hat{F}$ for the breadth-first ordering of the slots of the contention tree.

To be more formal about the exit times we introduce an ordering, $\leq$, on the elements in $\mathcal{J}$, and $\leq$ orders first on the length of the string and then lexicographically the strings with the same length. Use $\mathcal{J}_+$ to denote the set of all strings of positive arbitrary length: $\mathcal{J}_+ := \mathcal{J} \setminus \emptyset$. For $v \in \mathcal{J}_+$, define $v_- = u$ if $v \in u_+$. Next, for given $u \in \mathcal{J}$, let $T(u)$ denote the cumulative number of slots actually in the tree, and let $S(u)$ denote the cumulative number of exits from the tree, i.e.

$$T(u) := \sum_{v \in \mathcal{J}_+ : v \leq u} \mathbf{1}(N(v_-) > 1), \qquad (3.37)$$

and

$$S(u) := \sum_{v \in \mathcal{J}_+ : v \leq u} \mathbf{1}(N(v_-) > 1, N(v) = 1). \qquad (3.38)$$

Note that $S(\emptyset) = T(\emptyset) = 0$ by definition.

The variable $T(u)$ can be interpreted as time, as it counts the number of slots that are actually used in the contention tree. Hence, a graph of $S(u)$ vs. $T(u)$, where u ranges over the elements in $\mathcal{J}$ in the appropriate order, defines a stochastic process. This process starts at 0 at time 0, and increases to n during the time needed to complete the tree. The process stops at time T, with T the length of the tree defined in (3.25).

Clearly, we can define this process for both breadth-first and depth-first ordering of the slots in the tree. These processes are illustrated in Fig. 3.4.

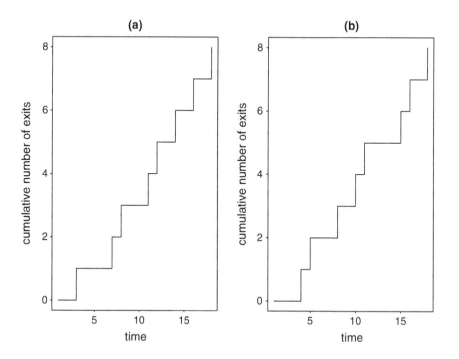

Fig. 3.4. Exit process from tree presented in Fig. 3.1 (**a**) for breadth-first order (**b**) for depth-first order

Figure 3.4a shows this process for the tree given in Fig. 3.1 with the breadth-first order and Fig. 3.4b shows the process for the tree given in Fig. 3.1 with the depth-first order. In the sequel, we limit ourselves to breadth-first trees.

This process also codes the empirical distribution, $\hat{F}$, of the success instants. To see this, let t denote a time on the positive time axis; let $u = T^{-1}(t)$ denote the node in $\mathcal{J}$ for which $T(u) \approx t$, i.e.

$$T^{-1}(t) = \begin{cases} \inf(u \in \mathcal{J} : T(u) = \lfloor t \rfloor) & \text{if } t \leq T, \\ \inf(u \in \mathcal{J} : T(u) = T) & \text{if } t > T, \end{cases} \tag{3.39}$$

then

$$\hat{F}(t) = \frac{1}{n} S(T^{-1}(t)). \tag{3.40}$$

It follows that the graph which interpolates the sequence of expected values of $T(u)$ and $S(u)$: $(\mathbb{E}(T(u)), \mathbb{E}(S(u)))$, provides a convenient approximation of the empirical distribution of the success instants.

The next theorem provides expressions for this distribution. Define $\mathbb{S}_d$ as the expected number of success instants up to level d, i.e. the expectation of (3.35) and $\mathbb{T}_d$ as the expected number of slots in the tree up to level d, i.e. the expectation of (3.33):

$$\mathbb{T}_d := \sum_{l=1}^{d} m^l q(l), \tag{3.41}$$

$$\mathbb{S}_d := \sum_{l=1}^{d} m^l q_1(l), \tag{3.42}$$

with q as defined in (3.8) and q_1 as defined in (3.14).

THEOREM 3.3 *With breadth-first order, the graph which interpolates the sequence* $(\mathbb{E}(T(u)), \mathbb{E}(S(u))), u \in \mathcal{J}$, *is given by*

$$F_b(t) := \frac{1}{n} \mathbb{S}_{d-1} + \frac{t - \mathbb{T}_{d-1}}{\mathbb{T}_d - \mathbb{T}_{d-1}} \frac{1}{n} (\mathbb{S}_d - \mathbb{S}_{d-1}), \text{ for } t \in (\mathbb{T}_{d-1}, \mathbb{T}_d]. \tag{3.43}$$

The proof of Theorem 3.3 is given in Sect. 3.4. The accuracy of the approximation depends on the variability of both $T(u)$ and $S(u)$. The following theorem states that the variability of both $T(u)/n$ and $S(u)/n$ is asymptotically negligible.

THEOREM 3.4 *For each* $u \in \mathcal{J}$, *under breadth-first order,*

$$\lim_{n \to \infty} \frac{1}{n^2} \text{var}(T(u)) = 0, \tag{3.44}$$

$$\lim_{n \to \infty} \frac{1}{n^2} \text{var}(S(u)) = 0. \tag{3.45}$$

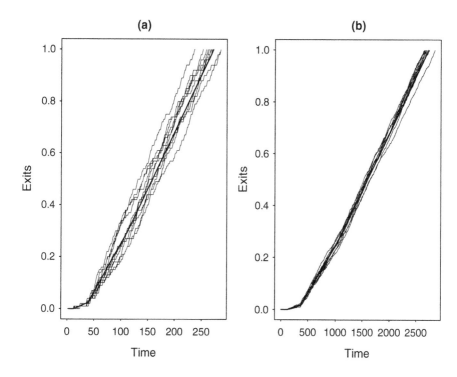

Fig. 3.5. Empirical cumulative distributions of success instants from 10 random contention trees, *light lines*, and theoretical approximation, *bold line*, for breadth-first order and (**a**) $n = 100$ (**b**) $n = 1,000$

Again, a proof is deferred to Sect. 3.4. In addition, we conjecture that similar theorems will hold for depth-first trees; see Sect. 3.5.

Theorem 3.4 shows that asymptotically, for $n \rightarrow \infty$, the empirical distribution function of the success instants normalised by n converges to its expected value. Theorems 3.3 and 3.4 are illustrated in Fig. 3.5. In Fig. 3.5, we have displayed the empirical distribution functions of the success instants for 10 random contention trees with breadth-first order, and the theoretical approximation given in Theorem 3.3. We have done so in Fig. 3.5a for $n = 100$ and in Fig. 3.5b for $n = 1,000$. We see that the theoretical approximation is quite good for both cases, and that the difference between the empirical distributions and theoretical distribution function becomes smaller for increasing n.

Theorems 3.3 and 3.4 do not imply, however, that the sequence of expected values itself converges to any fixed function. Rather, the expected value of the sequence of distribution functions will vary with ξ as $n = \xi m^k$ tends to infinity via $k \rightarrow \infty$. This is similar to the phenomenon that was observed for the expected tree length and the variance of the tree length considered in Sect. 3.3.1. A precise statement is given in Theorem 3.5.

THEOREM 3.5 *Let* $X \sim F_b$, *with* F_b *defined in* (3.43). *Then, for* $n = \xi m^k$ *with* $\xi \in [1, m)$,

$$\lim_{k \to \infty} \mathbb{P}\left(\frac{X}{\xi m^k} \leq t\right) = F_b^{\xi}(t), \tag{3.46}$$

where F_b^{ξ} *is defined for* $t \in (\mathbb{T}_{\delta-1}(\xi), \mathbb{T}_{\delta}(\xi)]$ *as*

$$F_b^{\xi}(t) := e^{-\xi m^{-\delta}} + \frac{t - \mathbb{T}_{\delta-1}(\xi)}{mh(\xi m^{-\delta+1})}\left(e^{-\xi m^{-\delta-1}} - e^{-\xi m^{-\delta}}\right), \tag{3.47}$$

with $\delta \in \mathbb{Z}$,

$$\mathbb{T}_{\delta}(\xi) = m \sum_{i=-\infty}^{\delta} h(\xi m^{-i}), \tag{3.48}$$

and h *as in* (3.28).

The proof uses (3.23) in Remark 3.3 and the fact that

$$\frac{\mathbb{S}_d}{n} = \left(1 - \frac{1}{m^d}\right)^{n-1}, \tag{3.49}$$

and is not further considered. Theorem 3.5 is illustrated in Fig. 3.6a,b. There, we have displayed the asymptotic forms of the distribution functions for

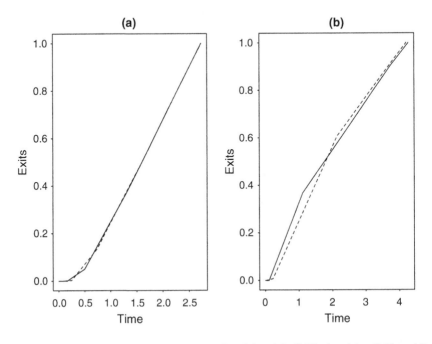

Fig. 3.6. Asymptotic distribution (**a**) for $m = 3$ and $\xi = 1$ (*solid line*) and $\xi = 2$ (*dotted line*) (**b**) for $m = 10$ and $\xi = 1$ (*solid line*) and $\xi = 5$ (*dotted line*)

various values of ξ. It can be observed that the distribution functions vary with ξ. It can also be observed that the oscillations with $m = 10$ are much larger than the oscillations with $m = 3$.

3.3.3 Skip Levels

From Fig. 3.5, we observe that the probability of success is very small during the first slots of the tree. Thus, for a tree with 100 contenders, the frequency of successes is very low during the first 50 slots of the tree, corresponding to the first three levels of the tree. For a tree with 1,000 contenders, the frequency of successes is very low up to level five.

These initial slots are essentially wasted. They only serve to split the initial group into smaller subgroups. However, this can also be achieved by starting the tree at a higher level. This hints at a method for optimising the tree by skipping the initial levels of the tree in which no successes occur, and then proceeding as usual. This idea is illustrated in Figs. 3.7 and 3.8. The tree in Fig. 3.7 is a standard contention tree. Note that there are no success instants during the first level of the tree. Skipping this level, as done in the tree in Fig. 3.8, thus decreases the total number of slots needed to resolve all the conflicts and increases the efficiency of the tree.

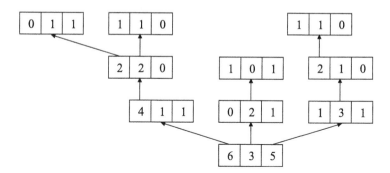

Fig. 3.7. Contention tree

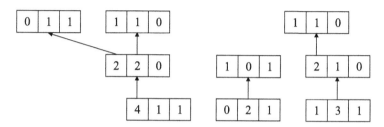

Fig. 3.8. Same contention tree as in Fig. 3.7, now started at the 2nd level

Generally, of course, there is no way of knowing in advance that no success instant will occur in a given level of the tree. However, by skipping initial levels in which success instants are sufficiently unlikely a priori, we can achieve the same effect and increase the efficiency of the tree. This has been explored in e.g. Capetanakis [30] and Mathys and Flajolet [120].

We now make the use of skip level trees more precise. Use $\mathbb{T}^{(s)}$ for the expected length of a tree that starts at level s. Adapting the approach from Sect. 3.3.1, it is easily seen that the expression for the expected tree length now becomes

$$\mathbb{T}^{(s)} = m^s + \sum_{i=s+1}^{\infty} m^i q(i), \qquad (3.50)$$

rather than (3.26). Using similar techniques as in Sect. 3.3.1, we obtain that

$$\lim_{k\to\infty} \frac{1}{n}\mathbb{T}^{(s)} = \lim_{k\to\infty} \frac{m^{s-k}}{\xi} + m \sum_{i=s-k}^{\infty} h(\xi m^{-i}), \qquad (3.51)$$

where $n = \xi m^k$ and where the dependence of n on k has been suppressed as usual.

The optimal number of levels to skip can now be computed by minimising (3.50) over s. We will not consider this optimisation in great detail. However, from (3.51), it is clear that we should choose $s = k + b$ for some constant b, and with some experimentation, we can numerically optimise for b. It appears that we should take $b \approx -1$. Thus, we recover the well known fact that we should start the tree with g^* subgroups, where g^* is chosen so that the expected number of contenders in each subgroup is slightly greater than 1, see, e.g. Bertsekas and Gallager [14], p. 291. Our approximation here is rather crude, as we have restricted ourselves to initial subgroups of size $g = m^s$.

Table 3.3 provides the optimal s for various values of n and $m = 3$. Moreover, it gives the average length of trees that start at this optimal level, and, for comparison, the average length of trees that start at level 1. Apparently, we can reduce the length of the contention tree by approximately 10%.

Table 3.3. Optimal starting level s_{opt}, average length of usual contention tree, $\mathbb{T}^{(1)}$, and average length of contention tree that starts at level s_{opt}, $\mathbb{T}^{(s_{opt})}$, for various values of n

n	s_{opt}	$\mathbb{T}^{(1)}$	$\mathbb{T}^{(s_{opt})}$
50	2	132	123
100	3	271	241
1,000	5	2,729	2,426

3.4　Proofs

Proof of Lemma 3.2. (a) We proceed using induction. For $d = 1$, the statement is obvious. Assuming that the statement is true up to level $d - 1$, for $z = \langle z_1 \ldots z_d \rangle$,

$$\mathbb{P}(N(z) = i) = \sum_{l=i}^{n} \mathbb{P}(N(z) = i | N(z_-) = l) \mathbb{P}(N(z_-) = l).$$

Next, use the definition of $N(z) | N(z_-)$ and the induction hypothesis to show (3.5).

(b) Use $\mathbb{P}(N(u) = i, N(v) = j) = \mathbb{P}(N(u) = i | N(v) = j) \mathbb{P}(N(v) = j)$ and (a).

(c) Statement (c) is clear if $u_1 \neq v_1$ so that u and v have no common ancestor. Next, assume that u and v have a latest common ancestor w at level f, i.e. $w_i = u_i = v_i, i = 1, \ldots, f$, and $u_{f+1} \neq v_{f+1}$. In this case,

$$\mathbb{P}(N(u) = i, N(v) = j)$$
$$= \sum_{l=i+j}^{n} \mathbb{P}(N(u) = i, N(v) = j | N(w) = l) \mathbb{P}(N(w) = l).$$

Now use (a) and (b) and evaluate the sum to obtain (3.7). □

Proof of Lemma 3.3. (a) Note that $\mathbb{P}(\tau(x) \geq d) = \mathbb{P}(N(x(d-1)) > 1)$, so that (a) follows from Lemma 3.2(a).

(b) Note that

$$\mathbb{P}(\tau(x) \geq d, \tau(y) \geq e) = \mathbb{P}(N(x(d-1)) > 1, N(y(e-1)) > 1)$$
$$= \mathbb{P}(N(y(e-1)) > 1),$$

for $x(d-1) \sim y(e-1)$ with $1 \leq d \leq e$. Hence, (b) follows from (a) in Lemma 3.2.

(c) Note that

$$\mathbb{P}(\tau(x) \geq d, \tau(y) \geq e)$$
$$= \mathbb{P}(N(x(d-1)) > 1, N(y(e-1)) > 1)$$
$$= 1 - \mathbb{P}(N(x(d-1)) \leq 1) - \mathbb{P}(N(y(e-1)) \leq 1)$$
$$+ \mathbb{P}(N(x(d-1)) \leq 1, N(y(e-1)) \leq 1),$$

and (c) follows from Lemma 3.2(a) and (c). □

Proof of Lemma 3.4. (a) Observe that

$$\mathbb{P}(I_1(x, d)) = \mathbb{P}(\tau(x) = d, \psi(x) = 1)$$
$$= \mathbb{P}(N(x(d-1)) > 1, N(x(d)) = 1),$$

so that the statement follows from Lemma 3.2(b):

$$
\begin{aligned}
\mathbb{P}(I_1(x,d)) &= \sum_{i=2}^{n} \binom{n}{1, i-1, n-1} \frac{1}{m^d} \left(\frac{m-1}{m^d}\right)^{i-1} \left(1 - \frac{1}{m^{d-1}}\right)^{n-i} \\
&= \frac{n}{m^d} \sum_{i=1}^{n-1} \binom{n-1}{i} \left(\frac{m-1}{m^d}\right)^{i} \left(1 - \frac{1}{m^{d-1}}\right)^{n-1-i} \\
&= \frac{n}{m^d} \left(\left(1 - \frac{1}{m^d}\right)^{n-1} - \left(1 - \frac{1}{m^{d-1}}\right)^{n-1} \right).
\end{aligned}
$$

(b) In this case

$$
\begin{aligned}
\mathbb{P}(I_1(x,d), I_1(y,d)) \\
= \mathbb{P}(N(x(d-1)) > 1, N(x(d)) = 1, N(y(e-1)) > 1, N(y(e)) = 1).
\end{aligned}
$$

However, as $x(d) \sim y(e)$ with $1 \le d \le e$, it follows that $x(d) \sim y(e-1)$. Hence, $N(y(e-1)) > 1$ contradicts $N(x(d)) = 1$ unless $d = e$. In the latter case (a) applies.

(c) Observe that $x(d-1) \sim y(d-1)$ implies that $x(d-1) = y(d-1)$. Hence, for x, y such that $x(d-1) \sim y(d-1)$ and $x(d) \not\sim y(d)$, we have

$$
\begin{aligned}
&\mathbb{P}(I_1(x,d), I_1(y,d)) \\
&= \mathbb{P}(N(x(d-1)) > 1, N(y(d-1)) > 1, N(x(d)) = 1, N(y(d)) = 1) \\
&= \mathbb{P}(N(x(d-1)) > 1, N(x(d)) = 1, N(y(d)) = 1) \\
&= \sum_{i=2}^{n} \mathbb{P}(N(x(d))=1, N(y(d)) = 1 | N(x(d-1)) = i)\mathbb{P}(N(x(d-1)) = i).
\end{aligned}
$$

Hence, on using Lemma 3.2(a) and (c), (3.18) follows.

(d) For x, y such that $x(d-1) \sim y(e-1)$ and $x(d) \not\sim y(e)$, with $1 \le d < e$, we have

$$
\begin{aligned}
\mathbb{P}(I_1(x,d), I_1(y,e)) = \sum_{2 \le i < j \le n} &\mathbb{P}(N(y(e)) = 1 | N(y(e-1)) = i) \\
&\times \mathbb{P}(N(x(d-1)) = j, N(y(e-1)) = i, N(x(d)) = 1).
\end{aligned}
$$

Moreover, observe that

$$
c_2(d,e) := \mathbb{P}(N(x(d-1)) = j, N(y(e-1)) = i, N(x(d)) = 1)
$$

reduces to

$$
\begin{aligned}
c_2(d, e) &= \mathbb{P}(N(y(e-1)) = i, N(x(d)) = 1 | N(x(d-1)) = j) \\
&\quad \times \mathbb{P}(N(x(d-1)) = j) \\
&= \binom{j}{1, i, j-i-1} \left(\frac{1}{m}\right) \left(\frac{1}{m^{e-d}}\right)^i \left(1 - \frac{1}{m} - \frac{1}{m^{e-d}}\right)^{j-i-1} \\
&\quad \times \mathbb{P}(N(x(d-1)) = j),
\end{aligned}
$$

where we have used Lemma 3.2(c).

Now putting this together, we obtain that

$$
\begin{aligned}
\mathbb{P}(I_1(x, d), I_1(y, e)) &= \sum_{j=3}^{n} \sum_{i=2}^{j-1} \binom{i}{1} \left(\frac{1}{m}\right) \left(1 - \frac{1}{m}\right)^{i-1} \binom{j}{1, i, j-i-1} \\
&\quad \times \left(\frac{1}{m}\right) \left(\frac{1}{m^{e-d}}\right)^i \left(1 - \frac{1}{m} - \frac{1}{m^{e-d}}\right)^{j-i-1} \\
&\quad \times \binom{n}{j} \left(\frac{1}{m^{d-1}}\right)^j \left(1 - \frac{1}{m^{d-1}}\right)^{n-j},
\end{aligned}
$$

and the desired result (3.19) follows after some more elementary manipulations.

(e) There is the elementary identity

$$
\begin{aligned}
f(a, b, c) &:= \sum_{i+j \le n} ij \binom{n}{i, j, n-i-j} a^i b^j c^{n-i-j} \\
&= n(n-1)ab(a+b+c)^{n-2}.
\end{aligned}
$$

Define

$$
h(a, b, c) := \sum_{\substack{i+j \le n \\ i,j \ge 2}} ij \binom{n}{i, j, n-i-j} a^i b^j c^{n-i-j}, \tag{3.52}
$$

and note that

$$
\begin{aligned}
h(a, b, c) &= f(a, b, c) - \sum_{i=1}^{n-1} i \binom{n}{i, 1, n-i-1} a^i b c^{n-i-1} \\
&\quad - \sum_{j=1}^{n-1} j \binom{n}{1, j, n-1-j} a b^j c^{n-1-j} + \binom{n}{1, 1, n-2} a b c^{n-2} \\
&= n(n-1)ab(a+b+c)^{n-2} + n(n-1)abc^{n-2} \\
&\quad - n(n-1)ab(b+c)^{n-2} - n(n-1)ab(a+c)^{n-2}.
\end{aligned}
$$

Observe that

$$
\mathbb{P}(I_1(x,d), I_1(y,e)) = \sum_{\substack{i+j\leq n \\ i,j\geq 2}} \mathbb{P}(N(x(d)) = 1 | N(x(d-1)) = i)
$$
$$
\times \mathbb{P}(N(y(e)) = 1 | N(y(e-1)) = j)
$$
$$
\times \mathbb{P}(N(x(d-1)) = i, N(y(e-1)) = j).
$$

Here the univariate conditional probabilities are simple binomial probabilities, and the bivariate probabilities are given in Lemma 3.2(c). Substitution yields

$$
\mathbb{P}(I_1(x,d), I_1(y,e)) = \frac{1}{(m-1)^2} \sum_{\substack{i+j\leq n \\ i,j\geq 2}} ij \binom{n}{i,j,n-i-j}
$$
$$
\times \left(\frac{m-1}{m^e}\right)^i \left(\frac{m-1}{m^d}\right)^j
$$
$$
\times \left(1 - \frac{1}{m^{d-1}} - \frac{1}{m^{e-1}}\right)^{n-i-j}
$$
$$
= \frac{1}{(m-1)^2} h\left(\frac{m-1}{m^e}, \frac{m-1}{m^d}, 1 - \frac{1}{m^{d-1}} - \frac{1}{m^{e-1}}\right),
$$

with h as defined in (3.52). Equation (3.20) now immediately follows. □

Proof of Theorem 3.3. Consider $u = \langle u_1 \ldots u_d \rangle$. From (3.37) we obtain that

$$
\mathbb{E}(T(u)) = \mathbb{E}\left(\sum_{v\in\mathcal{J}_+:v\leq u} \mathbf{1}(N(v_-) > 1)\right)
$$
$$
= \mathbb{E}\left(\sum_{l=1}^{d} \sum_{v\in\mathcal{J}_l:v\leq u} \mathbf{1}(N(v_-) > 1)\right)
$$
$$
= \sum_{l=1}^{d-1} m^l \mathbb{P}(\tau \geq l) + \mathbb{E}\left(\sum_{v\in\mathcal{J}_d:v\leq u} \mathbf{1}(N(v_-) > 1)\right)
$$
$$
= \mathbb{T}_{d-1} + n_d(u)q(d), \tag{3.53}
$$

where $n_d(u)$ is the number of slots at level d that precede u:

$$
n_d(u) = \sum_{i=1}^{d} (u_i - 1) n^{d-i} + 1. \tag{3.54}
$$

Similarly, we can evaluate $\mathbb{E}(S(u))$:

$$
\mathbb{E}(S(u)) = \mathbb{S}_{d-1} + n_d(u)q_1(d). \tag{3.55}
$$

Solving for $n_d(u)$ from (3.53) and inserting in (3.55), we obtain

$$
\begin{aligned}
\mathbb{E}(S(u)) &= \mathbb{S}_{d-1} + \frac{\mathbb{E}(T(u)) - \mathbb{T}_{d-1}}{q(d)} q_1(d) \\
&= \mathbb{S}_{d-1} + \frac{\mathbb{E}(T(u)) - \mathbb{T}_{d-1}}{m^d q(d)} m^d q_1(d) \\
&= \mathbb{S}_{d-1} + \frac{\mathbb{E}(T(u)) - \mathbb{T}_{d-1}}{\mathbb{T}_d - \mathbb{T}_{d-1}} (\mathbb{S}_d - \mathbb{S}_{d-1}),
\end{aligned}
$$

and (3.43) follows. $\qquad\square$

Proof of (3.44) in Theorem 3.4 concerning $\mathrm{var}(T(u))$. We consider $T(u)$ as defined in (3.37) and will evaluate $\mathrm{var}(T(u))$. First observe that to prove Theorem 3.4, we can restrict ourselves to $u = \langle u_1 \dots u_k\, m \rangle$. Next, we split $T(u)$:

$$
T(u) = \sum_{v \in \mathcal{J}_+ : v \leq u} \mathbf{1}(N(v_-) > 1) = \sum_{d=1}^{k+1} m^d \sum_{x \in \mathcal{X} : x(d) \leq u} \mathbf{1}(\tau(x) \geq d)\mathrm{U}(x)
$$

$$
= T_k + m^{k+1} \sum_{x \in \mathcal{X} : x(k+1) \leq u} \mathbf{1}(\tau(x) \geq k+1)\mathrm{U}(x).
$$

Now by an argument entirely analogous to the one leading to the expression for $\mathrm{var}(T)$ in (3.31), we find that

$$
\mathrm{var}(T(u)) = \mathbb{E}(T(u)^2) - (\mathbb{E}T(u))^2
$$

$$
= \sum_{d=1}^{k} ((q(d) - q(d)^2)m^{d+1} + (n(d, d) - q(d)^2)m^{d+1}(m^{d-1} - 1))
$$

$$
+ 2\sum_{d=1}^{k-1}\sum_{e=d+1}^{k} (q(e) - q(d)q(e))m^{e+1}
$$

$$
+ 2\sum_{d=1}^{k-1}\sum_{e=d+1}^{k} (n(d, e) - q(d)q(e))(m^{d+e} - m^{e+1}) + r(u).
$$

Here, $r(u)$ is a remainder term, relative to Expression (3.31), due to the inclusion of incomplete rows in the tree expansion defined via $T(u)$. To give an expression for $r(u)$, define $f(u)$ for the fraction of paths $x \in \mathcal{X}$ such that $x(k+1) \leq u$:

$$
f(u) := n_{k+1}(u)/m^{k+1}, \tag{3.56}
$$

with $n_d(u)$ as defined in (3.54). With this notation, $r(u)$ becomes

$$
\begin{aligned}
r(u) \;=\; & f(u)^2(q(k+1) - q(k+1)^2)m^{k+2} \\
& + (n(k+1, k+1) - q(k+1)^2)m^{k+2}(m^k - 1) \\
& + 2f(u)\sum_{d=1}^{k}(q(k+1) - q(d)q(k+1))m^{k+2} \\
& + 2f(u)\sum_{d=1}^{k}(n(d, k+1) - q(d)q(k+1))(m^{d+k+1} - m^{k+2}).
\end{aligned}
$$

Hence, in parallel with Theorem 3.2, we see that $\operatorname{var}(T(u))/n$ is bounded for any u, and (3.44) in Theorem 3.4 follows. $\qquad\square$

Proof of (3.45) in Theorem 3.4 concerning $\operatorname{var}(S(u))$. We first consider S_k as defined in (3.35), and find an expression for $\operatorname{var}(S_k)$. Note that

$$
\begin{aligned}
S_k^2 \;=\; & \sum_{d=1}^{k}\sum_{e=1}^{k} m^{d+e} \sum_{x,y\in\mathcal{X} } \mathbf{1}(I_1(x,d), I_1(y,d))\mathrm{U}(x,y) \\
\;=\; & \sum_{d=1}^{k} m^{2d} \sum_{x,y\in\mathcal{X}} \mathbf{1}(I_1(x,d), I_1(y,d))\mathrm{U}(x,y) \\
& + \sum_{d=1}^{k}\sum_{e=1}^{k} m^{d+e} \sum_{x,y\in\mathcal{X}} \mathbf{1}(I_1(x,d), I_1(y,e))\mathrm{U}(x,y).
\end{aligned}
$$

Now splitting this expression for S_k^2 into paths such that either $x(d)\sim y(e)$, $x(d)\nsim y(e)$ and $x(d-1)\sim y(e-1)$, or finally $x(d-1)\nsim y(e-1)$, as suggested by Lemma 3.4, and taking expectations, we get

$$
\begin{aligned}
\mathbb{E}(S_k^2) = & \sum_{d=1}^{k}\Big(m^d q_1(d) + (m^{d+1} - m^d)c_0(d) + (m^{2d} - m^d)n_1(d, d)\Big) \\
& + 2\sum_{d=1}^{k}\sum_{e=d+1}^{k}\Big((m^{e+1} - m^e)c_1(d, e) + (m^{d+e} - m^{e+1})n_1(d, e)\Big).
\end{aligned}
$$

We consider the sums over d and e and observe that the sum over e can be solved. In particular we find that

$$
\begin{aligned}
t_1(d) := & \sum_{e=d+1}^{k} (m^{e+1} - m^e)c_1(d, e) \\
= & \; n(n-1)\left(\frac{m-1}{m^d}\right)\left(\left(1 - \frac{1}{m^d} - \frac{1}{m^k}\right)^{n-2} - \left(1 - \frac{2}{m^d}\right)^{n-2}\right),
\end{aligned}
$$

and for the second term that

$$
\begin{aligned}
t_2(d) \;&:=\; \sum_{e=d+1}^{k} (m^{d+e} - m^{e+1}) n_1(d,e) \\
&=\; n(n-1)\left(1 - \frac{m}{m^d}\right)\left(\left(1 - \frac{m+1}{m^d}\right)^{n-2} - \left(1 - \frac{2}{m^d}\right)^{n-2}\right. \\
&\qquad \left. - \left(1 - \frac{1}{m^{d-1}} - \frac{1}{m^k}\right)^{n-2} + \left(1 - \frac{1}{m^d} - \frac{1}{m^k}\right)^{n-2}\right).
\end{aligned}
$$

Hence,

$$
\mathbb{E}(S_k^2) = \sum_{d=1}^{k} t(d),
$$

where

$$
\begin{aligned}
t(d) \;=\;& m^d q_1(d) + (m^{d+1} - m^d) c_0(d) + (m^{2d} - m^d) n_1(d,d) \\
&+ 2t_1(d) + 2t_2(d).
\end{aligned}
$$

Substitution of q_1, c_0, n_1, t_1, and t_2 in $t(d)$ and sorting the terms gives

$$
\begin{aligned}
t(d) \;=\;& n\left(\left(1 - \frac{1}{m^d}\right)^{n-1} - \left(1 - \frac{1}{m^{d-1}}\right)^{n-1}\right) \\
&+ n(n-1)\left(1 - \frac{1}{m^d} - \frac{1}{m^k}\right)^{n-2} 2\left(1 - \frac{1}{m^d}\right) \\
&+ n(n-1)\left(1 - \frac{1}{m^{d-1}} - \frac{1}{m^k}\right)^{n-2} 2\left(-1 + \frac{m}{m^d}\right) \\
&+ n(n-1)\left(1 - \frac{2}{m^d}\right)^{n-2}\left(-1 + \frac{1}{m^d}\right) \\
&+ n(n-1)\left(1 - \frac{2}{m^{d-1}}\right)^{n-2}\left(1 - \frac{m}{m^d}\right).
\end{aligned}
$$

Finally, summing over d, we find

$$
\mathbb{E}(S_k^2) = n\left(1 - \frac{1}{m^k}\right)^{n-2} + n(n-1)\left(1 - \frac{2}{m^k}\right)^{n-2}\left(1 - \frac{1}{m^k}\right). \tag{3.57}
$$

REMARK 3.4 *Observe that by inserting $k = \infty$ we have recovered the elementary fact that $\mathbb{E}(S)^2 = n^2$, so that indeed $\mathrm{var}(S) = 0$ as claimed in Sect. 3.3.1.*

REMARK 3.5 *The fact that the summation over d and e can be analytically carried out in the evaluation of $\mathbb{E}(S_k^2)$ should cause no surprise. Indeed, the* *number of successes up to level k equals the number of entries with a value equal to 1 at level k in the fully expanded complete m-ary tree, so that*

$$S_k = \sum_{u \in \mathcal{J}_k} \mathbb{1}(N(u) = 1). \tag{3.58}$$

Direct use of (3.58) rather than (3.35) leads to an alternative calculation of $\mathrm{var}(S_k)$.

A similar simplification can be carried out for $(\mathbb{E}S_k)^2$ and yields

$$(\mathbb{E}S_k)^2 = n^2 \left(1 - \frac{1}{m^k}\right)^{2(n-1)}, \tag{3.59}$$

so that

$$\mathrm{var}(S_k) = n \left(\left(1 - \frac{1}{m^k}\right)^{n-2} - \left(1 - \frac{2}{m^k}\right)^{n-2} \left(1 - \frac{1}{m^k}\right) \right)$$
$$+ n^2 \left(\left(1 - \frac{2}{m^k}\right)^{n-2} \left(1 - \frac{1}{m^k}\right) - \left(1 - \frac{1}{m^k}\right)^{2(n-1)} \right). \tag{3.60}$$

Now by an argument, entirely parallel to the previous one, we find that

$$\mathrm{var}(S(u)) = \mathrm{var}(S_k) + f(u)^2[q_1(k+1)m^{k+1} + c_0(k+1)(m^{k+2} - m^{k+1})$$
$$+ n_1(k+1, k+1)(m^{2(k+1)} - m^{k+2}) - q_1^2(k+1)m^{2(k+1)}]$$
$$+ 2f(u) \sum_{d=1}^{k} \Big(c_1(d, k+1)(m^{k+2} - m^{k+1}) \tag{3.61}$$
$$+ n_1(d, k+1)(m^{d+k+1} - m^{k+2}) - q_1(d)q_1(k+1)m^{d+k+1} \Big),$$

where, again, $f(u)$ equals the fraction of paths $x \in \mathcal{X}$ such that $x(k+1) \le u$ given in (3.56). We now simplify the sums over d in (3.61). From elementary manipulations, it follows that

$$\mathrm{var}(S(u)) = \mathrm{var}(S_k)$$
$$+ f(u)^2[q_1(k+1)m^{k+1} + c_0(k+1)(m^{k+2} - m^{k+1})$$
$$+ n_1(k+1, k+1)(m^{2(k+1)} - m^{k+2}) - q_1^2(k+1)m^{2(k+1)}]$$
$$+ 2f(u) (s_1(k+1) - s_2(k+1)), \tag{3.62}$$

where s_1 is defined as

$$
\begin{aligned}
s_1(k+1) &= \sum_{d=1}^{k} c_1(d, k+1)(m^{k+2} - m^{k+1}) \\
&\quad + \sum_{d=1}^{k} n_1(d, k+1)(m^{d+k+1} - m^{k+2}) \\
&= n(n-1)\left(1 - \frac{1}{m^k}\right) \\
&\quad \times \left(\left(1 - \frac{1}{m^k} - \frac{1}{m^{k+1}}\right)^{n-2} - \left(1 - \frac{1}{m^k}\right)^{n-2}\right),
\end{aligned}
$$

and s_2 is defined as

$$
\begin{aligned}
s_2(k+1) &= \sum_{d=1}^{k} q_1(d) q_1(k+1) m^{d+k+1} \\
&= n^2\left(1 - \frac{1}{m^k} - \frac{1}{m^{k+1}} + \frac{1}{m^{2k+1}}\right)^{n-1} \\
&\quad - n^2\left(1 - \frac{2}{m^k} + \frac{1}{m^{2k}}\right)^{n-1}.
\end{aligned}
$$

We now turn to the claim (3.45) in Theorem 3.4. This follows from evaluating the limit for each of the three terms in (3.62). E.g. for $\mathrm{var}(S_k)/n^2$,

$$
\frac{1}{n^2}\mathrm{var}(S_k) \le \left((1 - \frac{2}{m^k})^{n-2}(1 - \frac{1}{m^k}) - (1 - \frac{1}{m^k})^{2(n-1)}\right) + \frac{1}{n}. \tag{3.63}
$$

Given arbitrary $\epsilon > 0$, there is an $n^* > 1/\epsilon$ such that for all $n \ge n^*$ and all $k \ge 1$,

$$
\left|\left(1 - \frac{2}{m^k}\right)^{n-2} - e^{-2\xi m^{-\delta}}\right| \le \epsilon, \tag{3.64}
$$

and

$$
\left|\left(1 - \frac{1}{m^k}\right)^{2(n-1)} - e^{-2\xi m^{-\delta}}\right| \le \epsilon, \tag{3.65}
$$

where ξ and δ are defined via $\nu = \lfloor \log_m(n) \rfloor$ as $\xi = nm^{-\nu}$ and $\delta = k - \nu$. Hence, for $n \ge n^*$, (3.63) can be bounded as

$$
\begin{aligned}
\frac{1}{n^2}\mathrm{var}(S_k) &\le 3\epsilon + \frac{1}{m^k}e^{-2\xi m^{-\delta}} \\
&= 3\epsilon + \frac{1}{n}\xi m^{-\delta}e^{-2\xi m^{-\delta}} \\
&\le 4\epsilon.
\end{aligned}
$$

Thus indeed $\lim \operatorname{var}(S_k)/n^2 = 0$ for all k in case that $n \to \infty$. The verification that the other terms in (3.62) also vanish in the limit requires the same type of argument.   □

3.5   Conclusion

In this chapter we have considered basic properties of contention trees. As a new result, we have derived an expression for the marginal distribution of the success instants when the tree is traversed in breadth-first order. This expression was shown to be asymptotically exact for an increasing number of contenders.

Various lines for further research suggest themselves. Firstly, Denteneer and Keane [54] derive expressions for the moments of this distribution in terms of infinite sums. It is relevant to investigate whether these delay moments have simple approximations, much as the expressions for the moments of the tree length given in Theorems 3.1 and 3.2 can be approximated by elementary expressions given in Janssen and de Jong [86].

Secondly, it is relevant to generalise the results on the delay in the breadth-first trees to delay in skip level trees and to delay in depth-first trees. In fact, in [54], we have shown that Theorem 3.3 has a counterpart for depth-first trees:

THEOREM 3.6 *With depth-first order,* $\mathbb{E}(T(u))$ *and* $\mathbb{E}(S(u))$ *are given by*

$$\mathbb{E}(T(u)) = \sum_{l=1}^{d} \frac{u_l - 1}{m^l} (\mathbb{T} - \mathbb{T}_{l-1}) + \sum_{l=1}^{d} q(l), \qquad (3.66)$$

$$\mathbb{E}(S(u)) = \sum_{l=1}^{d} \frac{u_l - 1}{m^l} (\mathbb{S} - \mathbb{S}_{l-1}) + \sum_{l=1}^{d} q_1(l), \qquad (3.67)$$

for $u = (u_1, \ldots, u_d)$, *where* q *is as defined in* (3.8) *and* q_1 *in* (3.14).

Figure 3.9 illustrates Theorem 3.6 and shows both empirical distribution functions obtained by simulation and the theoretical distribution defined via (3.66) and (3.67). Figure 3.9 suggests that Theorem 3.4 also holds for depth-first trees. Additionally, Fig. 3.9 suggests that F_d is asymptotically approximately linear. We conjecture that the distribution function for the success instants in skip level trees, with optimally chosen number of levels to skip, is also linear.

Thirdly, the results cited so far on the tree length have only considered moments. It is also relevant to investigate distributional issues, such as a central limit theorem for the length.

A fourth line of relevant research pertains to the delay distribution for contention trees in a dynamic environment in which new contenders arrive

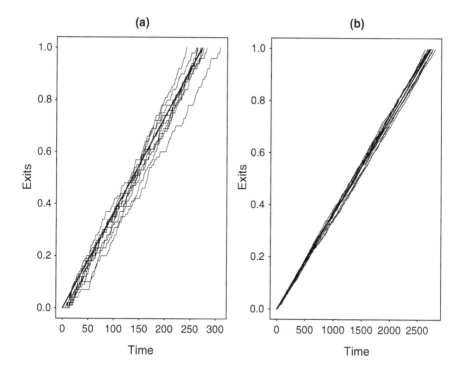

Fig. 3.9. Empirical cumulative distributions of success instants from 10 random contention trees, *light lines*, and theoretical approximation, *bold line*, for depth-first order: (**a**) $n = 100$ (**b**) $n = 1,000$

during the resolution of the conflicts in the current contention tree. Generally, the analysis of delay due to contention trees in such a dynamic environment is quite difficult, certainly for closed networks. This is due to the complicated nature of the distribution of the tree length, see e.g. Denteneer and Pronk [55]. The results from this chapter, however, point to some features of contention trees that help to simplify the problem. Particularly, note that linearity of both mean and variance of the length of contention trees naturally leads one to consider queueing approximations to contention trees. The usefulness of this idea is further corroborated by the linearity of the distribution of the success instants in case of depth-first and skip level trees, or by the almost linearity of this distribution for breadth-first trees. Such queueing approximations are the subject of the next chapter.

Chapter 4

DELAY MODELS FOR CONTENTION TREES IN CLOSED POPULATIONS

In this chapter we extend the study of stand-alone contention trees to more practical settings. In such a setting, one must take into account that stations alternate between activity periods and idle periods, and the splitting procedure must be complemented with a channel access protocol to accommodate newcomers. We describe the standard algorithms to do this, free access and blocked access, and propose a novel mechanism: Scheduled blocked access.

Motivated by the results in Sect. 1.4.1 and Chap. 3, we consider closed queueing models to approximate the delay due to contention resolution. More specifically, we study a number of variants of the standard repairman model, that differ in the service order at the repair facility. For each variant, we study the sojourn time at the repair facility. Moreover, we show that our results can be used to give excellent approximations to the request delay with contention trees in a finite population for any of the channel access protocols.

4.1 Introduction

In this chapter, we describe and model contention trees in cable networks, for which we must deal with the dynamics of the environment. These dynamics arise from the fact that users will alternate between idle periods and activity periods. During idle periods they have no need to access the shared resource, so that they do not participate in the contention process. During activity periods, however, a user wants to participate in the contention process in order to access the resource. It remains to decide how to accommodate newcomers who have just become active.

In this chapter, we consider three channel access protocols to do so. First, we describe the standard access methods: Free access and blocked access. Next, we introduce scheduled blocked access, which is a windowed access algorithm, see Sect. 2.1. All access methods are combined with the basic tree algorithms. The modified tree algorithm, see Sect. 2.1, is not considered.

A tractable model for the access delay due to contention trees when used in such a dynamic setting is an essential step towards a better understanding of the reservation procedure described in Sects. 1.2 and 1.3: Stations request data slots in contention with other stations via contention trees. After a successful request, data transfer follows in reserved slots, not in contention with other stations.

The analysis of the contention tree in Chap. 3 has revealed a number of properties:

- The average length of the tree is approximately proportional to the initial number of contenders, see Theorem 3.1.

- The variance of the length of the tree is, again approximately, proportional to the initial number of contenders in the tree, see Theorem 3.2.

- The success instants in the tree are, to a first approximation, uniformly distributed over the total time it takes to complete the tree. This is true, for a large number of contenders, for depth-first trees, see Fig. 3.9. For breadth-first trees, this observation holds to a lesser extent, see Fig. 3.5, as the tree has an initial phase in which no successes occur. It is only after this start-up phase that success instants are uniformly spaced.

These observations show that the departure process from a tree resembles a renewal process. Thus, they strongly motivate the use of queueing models to approximate contention trees in a dynamic setting. In these approximations, we view the shared channel as a central server that has to serve a stream of incoming jobs: The requests. The rate at which these jobs are served can be derived from the *capacity* of the contention tree, see Theorem 3.1 and (3.51) in Chap. 3. As we are looking for tractable finite-population models, see Sect. 1.4.1, the repairman model is a natural candidate model for the access delay in contention resolution with contention trees. In this chapter, we propose and analyse variants of the repairman model to obtain the required approximations.

The repairman model, see, e.g. Takács [159], Chap. 5, is also known as the computer terminal model or as the time sharing system. The basic model is illustrated in Fig. 4.1. There are N machines working in parallel. After a working period, a machine breaks down and joins the repair queue. At the repair facility, a single repairman repairs the machines according to some service discipline. Once repaired, a machine starts working again. We refer to Sect. 2.2 for a more extensive account of the repairman model.

In this chapter we show that the repairman model is an appropriate model for contention trees in a dynamic setting. The machines in the model correspond to

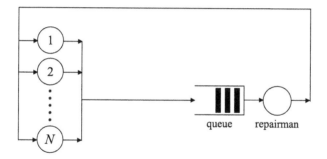

Fig. 4.1. Repairman model

the stations, the working periods of the machines correspond to the idle periods of the stations, and the time spent at the repair facility corresponds to the time spent in contention resolution.

It turns out that the average time spent in contention resolution, obtained via simulations, matches the average sojourn time at the repair facility in the basic repairman model with the First Come First Served (FCFS) discipline almost perfectly. However, the basic model fails to accurately predict the variance of the time spent in contention resolution.

Closer inspection of contention trees reveals a possible source for this mismatch. Contention trees operate by recursively splitting a group of stations into subgroups. Splitting stops as soon as each subgroup contains at most one station: A station is successful in transmitting its request as soon as it is the only contender in a group. The split is performed so that each station in a given group has the same probability of being successful, irrespective of the instant at which it became ready to transmit the request. Thus, contention trees differ from queues with a FCFS discipline. This suggests that variants of the basic repairman model are needed with some randomness built into their service discipline. In this chapter, we consider three such variants.

The first model that we consider is the repairman model with the Random Order of Service (ROS) discipline. This will serve as a model for the free access protocol. Here, after a repair, the next machine to be repaired is chosen randomly from the machines in the repair queue. We analyse the sojourn time distribution at the repair queue for this model using a connection with the repairman model considered in Mitra [126], in which the service discipline at the repair facility is Processor Sharing (PS).

Secondly, to model the blocked access protocol, we introduce an extension of the repairman model. In this extension, illustrated in Fig. 4.2, machines that break down are first gathered in an ante room before they are put in random order in the actual repair queue at the instants that the single server becomes idle. In the sequel this service discipline will be called Gated Random Order

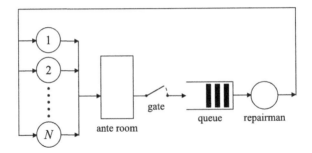

Fig. 4.2. Repairman model with gated random order of service

of Service (GROS). The emphasis of our analysis will be on obtaining an approximate expression for the variance of the sojourn time at the repair facility.

Thirdly, we turn to scheduled blocked access: See Denteneer [46] and [166]. Following Denteneer [44], this is modelled by means of a machine repair model, now with Gated Partial Random Order of Service (GPROS). This service discipline is a hybrid form of the FCFS and the GROS service disciplines. To explain GPROS in detail, we have to introduce the gate periods of the GROS discipline, which are periods between two successive openings of the gate. A gate period starts with the admission of a positive number of machines. Under the GPROS discipline, this gate period is divided into g intervals of equal length. At the start of a gate period, the arrivals during the previous gate period are placed in the queue so that all arrivals in the ith interval are before all arrivals in the intervals $i + 1, \ldots, g$. However, the relative order in which the arrivals within the ith interval are placed into the queue is random. This procedure can be thought of as a compromise between FCFS, take $g = \infty$, and GROS, take $g = 1$.

These different scheduling disciplines do not affect the mean sojourn time. This is a direct consequence of Little's formula and the fact that the number of customers in the single-server queue is the same for any work-conserving service discipline that does not pay attention to the actual service requests of the customers. However, the randomness does impact the higher order statistics and in particular the variance.

In our analysis of these models, we will therefore focus on approximations to the second moment of the access delay, and so it is appropriate to comment briefly on the relevance of the variance of the access delay in contention resolution. Firstly, low variability implies low jitter. As such, access variability is a key performance measure in itself. A second reason for studying the variance of the access delay is that it determines the variance of the request size in a reservation procedure, due to the request merging described in Sect. 1.3, which will be more extensively covered in Sect. 4.2. Thus, the variance of the access delay is needed in understanding the total average packet delay in cable networks: Both the average request delay per packet and the average packet transmission delay depend on the variance of the request size.

The rest of this chapter is organised as follows. In Sect. 4.2 we describe the contention-resolution process using contention trees in more detail. Next, in Sect. 4.3, we review some of the properties of the basic repairman model. Moreover, we derive expressions for the first two moments of the steady-state sojourn time distribution. The repairman model with ROS is considered in Sect. 4.4. We first relate the model with ROS to the model with PS. After that, we briefly review the main results from Mitra [126] for the model with PS. In Sect. 4.5, we give an approximate derivation of the moments of the sojourn time in the model with GROS and in Sect. 4.6, we do the same for GPROS. In Sect. 4.7 we present numerical results which show that the models of Sects. 4.4 – 4.6 can be used to approximate the request delay for contention resolution in a reservation procedure with contention trees. Finally, Sect. 4.8 presents some conclusions.

4.2   Access via Contention Trees

The basic splitting mechanism of the contention tree was described in Chap. 3. This algorithm must be complemented with a *channel access protocol* that specifies the procedure to be followed by stations that have data to transmit and that are not already contending in the tree. There are two basic access protocols: Free access and blocked access. In the former protocol, access to the tree is free and any station can transmit a request in the next node of the tree, as soon as it has data to transmit. In the latter protocol, the tree is blocked so that new stations can only transmit requests in the root node of the tree that is started as soon as the current tree has been completed.

The stations exhibit the following behaviour:

- A station becomes active in the contention process when it generates a data packet. In case of free access, it will then transmit a request in the next tree node, randomly choosing one of the three slots in this node. In case of blocked access, it will wait for the next new tree to be started and transmit its request in one of the slots of the root node of this tree. The request message includes a field called the request size, which indicates the number of data packets for which transfer is requested.

- The station stays active until its request has been successfully transmitted.

- While active, the station can update the request size included in the request message. Hence, packets that are generated at such an active station do not cause extra requests.

- After successful transmission of the request, the station quits the reservation process, to become active again when it generates a new data packet.

Note that request merging implies that the number of stations that can be active in contention is bounded. It is exactly this property that makes open models less appropriate to describe contention resolution for a reservation procedure. This property also explains the approach in this chapter, which approximates the request delay in transmitting a request by means of the sojourn time in a repairman model.

Next, observe that ROS serves as an approximation to the free access protocol: Neither in trees with free access nor in trees with ROS is there any advantage for contenders that are active over newly arriving contenders.

The GROS discipline serves as an approximation to blocked access. Indeed, both mechanisms have the concept of a batch. With GROS, it is the number of stations that enter in one gate period, and with the blocked contention trees, it is the group of stations that attempt to transmit in a new root node of the tree. Moreover, stations in the same batch are served in random order for both mechanisms.

4.2.1   Scheduled Blocked Access

We consider a variation on the blocked ternary tree with random splits. In this variation, the initial group of stations is split into g subgroups rather than three. Moreover, we assume that these initial subgroups are defined in terms of the instants at which the stations became active. More formally, we assume that the previous tree was executed in the time interval $[t_1, t_2]$. A station then chooses subgroup i in the initial split if it became active in the interval $[t_1 + (i-1)(t_2 - t_1)/g, t_1 + i(t_2 - t_1)/g]$, where i ranges from 1 to g. Contention in each of these g initial subgroups is then further organised according to a ternary tree as considered above, in which the splits are based on randomisation.

Thus we obtain a contention tree which is a hybrid of a ternary contention tree with random splits and a FCFS contention tree, see, e.g. Bertsekas and Gallager [14], Sect. 4.3.2 for a description of the latter. Such a hybrid form is appropriate in case the stations have a crude notion of time, which is sufficient to carry out an initial split into subgroups, but which is insufficiently accurate to carry out the full splitting.

Alternatively, this situation arises as an approximation to advanced tree schedules considered in Denteneer [46]; also described in [74, 166]. There are also similarities to windowed access algorithms considered in Mosely and Humblett [127] and in Van den Broek [26]. In such a tree schedule, new trees are started regularly and all stations that have data to transmit and that are not yet contending in another tree, will be the participants in this new tree. Hence there are multiple trees active at any instant. These trees are dealt with in a

FCFS manner, so that the contentions within the oldest tree are resolved first. However, within a tree, the order in which individual stations are successful is random and not related to the instants at which they became active.

Note that these scheduled blocked access trees improve upon the basic blocked tree in two respects. Firstly, one can vary the degree of the initial level of the tree with the traffic intensity. Thus one implements a skip level tree as described in Sect. 3.3.3. Such trees have an advantage over the basic contention tree in that fewer slots are required to complete the tree and this reduces average delay. Secondly, this mechanism introduces a negative correlation between the waiting time before the start of the tree and the waiting time within the tree. In this way, the variance of the total access delay is reduced as compared to the standard blocked contention tree. The relevance of the variance of the access delay was stressed in the introduction of this chapter.

In an actual implementation, one has to choose g, the initial number of sub-groups into which the contenders are to be split. A suitable way is to choose g so that the expected tree length is minimised. For the case of ternary trees, it was argued in Sect. 3.3.3 that one may take

$$g \approx \lfloor {}^3\!\log(n/3) \rfloor, \tag{4.1}$$

where n is the initial number of contenders in the tree. Of course, in applications such as cable networks, n is not known in advance, so that one must employ a procedure to estimate n, see, e.g. Denteneer [46] or Yin and Lin [175].

The GPROS model serves as an approximation to scheduled blocked access. Now, the stations that enter into slot i of the root node of the tree correspond to the arrivals during the ith of the intervals that make up a complete gate period of the GPROS model.

4.3 Properties of the Basic Model

First we introduce some notation and recall some properties of the basic repairman model; cf. Fig. 4.1. The total number of machines in the system is denoted by N. The machines work in parallel and break down, independently, after an exponentially distributed working period with parameter λ. Machines that break down join the repair queue, where they are served in a FCFS manner by a single repairman. The repair times are exponentially distributed with parameter μ. In Remark 4.1 we briefly comment on these distributional assumptions.

With the random variables X and Y we denote the steady-state number of machines that are working Q_W and that are in repair Q_R, respectively. Clearly, the number of working machines and the number of machines in repair evolve as Markov processes. Their steady-state distributions are given by, cf. Kleinrock [99] or Kobayashi [101],

$$\mathbb{P}(X = k) = \mathbb{P}(Y = N - k) = \frac{\rho^k/k!}{\sum_{i=0}^{N} \rho^i/i!}, \qquad k = 0, \ldots, N, \qquad (4.2)$$

where

$$\rho := \mu/\lambda. \qquad (4.3)$$

For the mean and variance of X and Y we have

$$\mathbb{E}(X) = \rho(1 - B_N(\rho)), \quad \mathbb{E}(Y) = N - \mathbb{E}(X), \qquad (4.4)$$
$$\text{var}(X) = \text{var}(Y) = \mathbb{E}(X) - \rho B_N(\rho)[N - \mathbb{E}(X)], \qquad (4.5)$$

where $B_N(\rho)$ denotes the Erlang loss probability, given by

$$B_N(\rho) = \frac{\rho^N/N!}{\sum_{i=0}^{N} \rho^i/i!}. \qquad (4.6)$$

Indeed, it is well known that the number of operative machines has the same distribution as the number of busy lines in the classical Erlang loss model.

We now turn to the moments of the sojourn time of an arbitrary machine at the repair facility. To this end, we consider a time epoch at which an arbitrary machine breaks down and moves to the repair queue. Stochastic quantities related to this instant are denoted by a subscript 1. Thus X_1 is the number of working machines at this instant, and Y_1 is the number of machines in repair at this instant. From the arrival theorem, see Sevcik and Mitrani [153], it follows that the distributions of X_1 and Y_1 are given by (4.2), but with N replaced by $N - 1$. We have for $k = 0, \ldots, N - 1$

$$\mathbb{P}(X_1 = k) = \mathbb{P}(Y_1 = N - 1 - k) = \frac{\rho^k/k!}{\sum_{i=0}^{N-1} \rho^i/i!}. \qquad (4.7)$$

The sojourn time of an arbitrary machine at the repair facility equals its own repair time plus the sum of the repair times of the machines already present at the repair facility. Thus, denoting this sojourn time by S, we have that

$$S = \sum_{i=1}^{Y_1+1} B_i, \qquad (4.8)$$

with $B_i, i = 1, 2, \ldots$, a sequence of independent, exponentially distributed random variables with parameter μ. Equation (4.8) enables us to obtain the Laplace–Stieltjes transform (LST) of the sojourn time at the repair facility, see also Kobayashi [101]:

$$\mathbb{E}(e^{-\omega S}) = \sum_{j=0}^{N-1} \frac{\rho^{N-1-j}/(N-1-j)!}{\sum_{i=0}^{N-1} \rho^i/i!} \left(\frac{\mu}{\mu+\omega}\right)^{j+1}. \qquad (4.9)$$

Here, we are mainly interested in the first two moments of the sojourn time. These can be obtained by consideration of the moments of the random sum (4.8), i.e.,

$$\mathbb{E}(S) = \mathbb{E}(Y_1 + 1)\mathbb{E}(B_1) = \frac{1}{\mu}\Big(N - \rho(1 - B_{N-1}(\rho))\Big), \qquad (4.10)$$

and

$$\begin{aligned} \mathrm{var}(S) &= \mathbb{E}(Y_1 + 1)\mathrm{var}(B_1) + \mathrm{var}(Y_1 + 1)\mathbb{E}(B_1)^2 \\ &= \frac{1}{\mu^2}\Big(N - \rho B_{N-1}(\rho)[N - 1 - \rho(1 - B_{N-1}(\rho))]\Big). \quad (4.11) \end{aligned}$$

Now, for N large and $N \gg \mu/\lambda$, $B_N(\rho)$ goes to zero like $\rho^N/N!$, and the following are extremely sharp approximations:

$$\mathbb{E}(S) \approx \frac{N}{\mu} - \frac{1}{\lambda}, \qquad (4.12)$$

$$\mathrm{var}(S) \approx \frac{N}{\mu^2}. \qquad (4.13)$$

In Sects. 4.4–4.6 we study the sojourn time distribution at Q_R for the ROS, GROS, and GPROS disciplines, respectively. As was briefly indicated in Sect. 4.1, the mean sojourn time at Q_R is the same under FCFS, ROS, GROS, and GPROS. Hence, we will concentrate on the variance of the sojourn time. Formula (4.13) shows that for the FCFS discipline, asymptotically, this variance is linear in the number of machines and does not depend on λ.

REMARK 4.1 It is well known that the repairman model has an insensitivity property: The steady-state distribution of the number of machines in repair only depends on the mean working time of machines and not on the actual distribution of these working times. This implies that the results of this section remain valid for arbitrarily distributed working times. Takács [159] has considered the situation with exponential working times, and general service time. The assumption of exponential service times is not very appropriate for application in the context of cable networks. However, we use the results for large N and $N\lambda > \mu$, in which case the dependence on the distributional assumptions is very weak.

REMARK 4.2 It is instructive to consider the situation in which N tends to infinity for fixed $\Lambda := N\lambda > \mu$. In this case, the considerations leading to (4.12) and (4.13) apply to (4.4) and (4.5), so that

$$\begin{aligned} \mathbb{E}(Y) &\approx N(1 - \mu/\Lambda), \\ \mathrm{var}(Y) &\approx N\mu/\Lambda. \end{aligned}$$

In this case μ/Λ is between 0 and 1. This suggests that, in this limiting regime, the variability in the number of machines in the repair queue is small as compared to the actual number of machines in the repair queue. It can be seen from (4.4) and (4.5) and the above, that in this case $\mathbb{E}(X) \approx N\mu/\Lambda$ and $\text{var}(X) \approx N\mu/\Lambda$, and indeed it can be seen from (4.2) that X here is asymptotically Poisson distributed.

4.4 ROS Discipline

Again we consider the model of Fig. 4.1, but now the service discipline at Q_R is ROS. For reasons that will soon become clear, we assume that the system contains $N+1$ rather than N machines. The main goals of this section are: (1) to determine the LST of the waiting time distribution at Q_R, (2) to relate this distribution to the sojourn time distribution at Q_R in case the service discipline is PS instead of ROS, and (3) to determine the asymptotic behaviour of the variance of the waiting (and sojourn) time at Q_R under the ROS discipline.

Consider a tagged machine, C, at the instant it arrives at Q_R. Let S_{ROS} (W_{ROS}) denote the steady-state sojourn (waiting) time of C at Q_R. S_{ROS} is the sum of W_{ROS} and a service time that is independent of W_{ROS}, and hence we can concentrate on W_{ROS}. We denote by $Y_1^{(N+1)}$ the number of machines in Q_R, as seen by C upon arrival at Q_R. Introduce for $j = 0, \ldots, N-1$,

$$\phi_j(\omega) := \mathbb{E}(e^{-\omega W_{ROS}} | Y_1^{(N+1)} = j+1), \quad \text{Re}(\omega) \geq 0,$$

where $\text{Re}(\omega)$ denotes the real part of ω. We can write, for $\text{Re}(\omega) \geq 0$,

$$\mathbb{E}(e^{-\omega W_{ROS}} | W_{ROS} > 0) = \sum_{j=0}^{N-1} \mathbb{P}(Y_1^{(N+1)} = j+1 | Y_1^{(N+1)} > 0)\phi_j(\omega).$$

$$(4.14)$$

For the N unknown functions $\phi_0(\omega), \ldots, \phi_{N-1}(\omega)$, the following set of N equations holds:

$$\phi_j(\omega) = \frac{\mu + (N-j-1)\lambda}{\mu + (N-j-1)\lambda + \omega} \left[\frac{(N-j-1)\lambda}{\mu + (N-j-1)\lambda}\phi_{j+1}(\omega) \right.$$
$$\left. + \frac{\mu}{\mu + (N-j-1)\lambda} \left(\frac{1}{j+1} + \frac{j}{j+1}\phi_{j-1}(\omega) \right) \right]. \quad (4.15)$$

Notice that the pre-factors of $\phi_{-1}(\omega)$ and $\phi_N(\omega)$ equal zero. Formula (4.15) can be understood in the following way. The pre-factor

$$\frac{\mu + (N-j-1)\lambda}{\mu + (N-j-1)\lambda + \omega}$$

is the LST of the time until the first *event*: Either an arrival at Q_R or a departure from Q_R. An arrival occurs first with probability

$$\frac{(N-j-1)\lambda}{\mu+(N-j-1)\lambda}.$$

In case of an arrival, the memoryless property of the exponential working and repair times implies that the tagged machine C sees the system as if it only now arrives at Q_R, meeting $j+2$ other machines there. A departure occurs first with probability

$$\frac{\mu}{\mu+(N-j-1)\lambda}.$$

In case of a departure, C is with probability $1/(j+1)$ the one to leave the ante room and enter the service position; if it does not leave, it sees Q_R as if it only now arrives, meeting j other machines there.

We can use (4.15) to obtain numerical values of $\mathbb{E}(W_{ROS}|W_{ROS} > 0)$ and $\mathrm{var}(W_{ROS}|W_{ROS} > 0)$, see Table 4.1. Formula (4.15) can also be used to study this mean and variance asymptotically, for $N \to \infty$. In fact, for this purpose we can also use the analysis given by Mitra [126] for a strongly related model: The machine-repair model with PS at Q_R and with N (instead of $N+1$) machines. Denote the LST of the sojourn time distribution of a machine meeting j machines at Q_R, in the case of PS, by $\psi_j(\omega)$. A careful study of Formula (4.15) reveals that exactly the same set of equations holds for $\psi_j(\omega)$, if in the PS case there are not $N+1$ but N machines in the system. If C meets j machines at the PS node Q_R, then it leaves $N-j-1$ machines behind at Q_W. Now observe that the time until either an arrival at or a departure from Q_R occurs is exponentially distributed with parameter $\mu+(N-j-1)\lambda$, leading to the same pre-factor as in (4.15). If an event occurs, it is a departure from Q_R with probability $\mu/(\mu+(N-j-1)\lambda)$. If a departure from Q_R occurs, C is with probability $1/(j+1)$ the machine to leave. If it does not leave, it sees Q_R as if it only now arrives, meeting $j-1$ machines there. Not only do we have

$$\phi_j(\omega) = \psi_j(\omega), \quad j = 0, \dots, N-1,$$

but it also follows from (4.7) that

$$\mathbb{P}(Y_1^{(N+1)} = j+1|Y_1^{(N+1)} > 0) = \mathbb{P}(Y_1^{(N)} = j), \quad j = 0, \dots, N-1.$$

The above equalities, combined with (4.14), imply that W_{ROS}, conditionally upon being positive, in the machine-repair system with $N+1$ machines, has the same distribution as the sojourn time under PS in the corresponding system with N machines. Adding a superscript (N) for the case of a machine-repair system with N machines, we can write

$$\mathbb{P}(S_{PS}^{(N)} > t) = \mathbb{P}(W_{ROS}^{(N+1)} > t|W_{ROS}^{(N+1)} > 0). \tag{4.16}$$

This equivalence result between ROS and PS may be viewed as a special case of a more general result in Borst et al. [21], see Cohen [37] for another special case. In the $G/M/1$ queue, the sojourn time under PS is equal in distribution to the waiting time under ROS of a customer arriving to a non-empty system [21]. This equivalence is extended in [21] to a class of closed product-form networks (notice that the two-queue network in the present chapter indeed is a closed product-form network). In particular, it follows from [21] that

$$
\begin{aligned}
\mathbb{P}(S_{PS}^{(N)} > t) &= \mathbb{P}(W_{ROS}^{(N+1)} > t | W_{ROS}^{(N+1)} > 0) \\
&= \frac{\mathbb{P}(W_{ROS}^{(N+1)} > t)}{\mathbb{P}(W_{ROS}^{(N+1)} > 0)}, \qquad t \geq 0, \tag{4.17}
\end{aligned}
$$

with

$$
\mathbb{P}(W_{ROS}^{(N+1)} > 0) = \frac{\sum_{i=0}^{N-1} \rho^i / i!}{\sum_{i=0}^{N} \rho^i / i!}. \tag{4.18}
$$

It is easily verified that, for the repairman model with N machines,

$$
\mathbb{E}(S_{ROS}) = \mathbb{E}(S_{PS}) = \mathbb{E}(S_{FCFS}),
$$

just as indicated in Sect. 4.1, the latter quantity equaling $\mathbb{E}(Y_1+1)/\mu$, cf. (4.10). For example, the first equality follows after some calculation from the following relation, that is obtained from (4.17) by integration over t:

$$
\mathbb{E}(S_{PS}^{(N)}) = \frac{\mathbb{E}(W_{ROS}^{(N+1)})}{\mathbb{P}(W_{ROS}^{(N+1)} > 0)}. \tag{4.19}
$$

Multiplication by t and integration over t in (4.17) similarly yields

$$
\mathrm{var}(S_{PS}^{(N)}) = \frac{\mathrm{var}(W_{ROS}^{(N+1)})}{\mathbb{P}(W_{ROS}^{(N+1)} > 0)}. \tag{4.20}
$$

If N is large and $N > \mu/\lambda$, then $\mathbb{P}(W_{ROS}^{(N+1)} = 0)$ is negligibly small. The previous formula hence implies that, for $N \to \infty$,

$$
\mathrm{var}(S_{PS}^{(N)}) \sim \mathrm{var}(W_{ROS}^{(N)}) \quad \text{and} \quad \mathrm{var}(S_{PS}^{(N)}) \sim \mathrm{var}(S_{ROS}^{(N)}).
$$

For an asymptotic analysis of $\mathbb{E}W_{ROS}^{(N)}$ and $\mathrm{var}(W_{ROS}^{(N)})$ we can thus immediately apply corresponding asymptotics of Mitra [126] for the PS-variant. Mitra [126] derives a similar set of equations, in matrix form, as (4.15), albeit for $\mathbb{P}(S_{PS} > t | Y_1 = j)$ rather than for its LST. He shows that the corresponding matrix has N real and negative eigenvalues $\mu_N \leq \mu_{N-1} \leq \cdots \leq \mu_1$. Using the equivalent of (4.14) for PS, he finally shows that

$$P(S_{PS} > u) = \sum_{i=1}^{N} \alpha_i e^{\mu_i u}, \qquad u \geq 0, \qquad (4.21)$$

with $\alpha_i > 0$ for $i = 1, \ldots, N$ and $\sum_{i=1}^{N} \alpha_i = 1$. Hence,

$$\mathbb{P}(S_{PS} > u) \leq e^{\mu_1 u}, \qquad (4.22)$$

$$\mathbb{P}(S_{PS} > u) \sim \alpha_1 e^{\mu_1 u}, \qquad u \to \infty. \qquad (4.23)$$

The fact that S_{PS} is hyper-exponentially distributed immediately implies that, see Proposition 12 in [126],

$$\mathrm{var}(S_{PS}) \geq \mathbb{E}(S_{PS})^2. \qquad (4.24)$$

Hence $\mathrm{var}(S_{PS}) = \mathcal{O}(N^2)$ for $N \to \infty$, which sharply contrasts with the $\mathcal{O}(N)$ behaviour for FCFS, see (4.13).

In Table 4.1, we consider $\mu = 0.5, 1$ and 2, giving rise to $\rho = \frac{1}{2}N$, $\rho = N$ and $\rho = 2N$, respectively. The next three remarks relate to these three different cases.

REMARK 4.3 Interestingly, in case $\rho = \mu/\lambda \ll N$, the standard deviation σ_{ROS} of the sojourn time for ROS is almost identical to $\mathbb{E}S = \mathbb{E}S_{ROS}$. This can be observed from the entry corresponding to $\mu = 0.5$ in Table 4.1. This suggests that for the considered parameter values, S_{ROS} is approximately exponentially distributed. Indeed, the following reasoning shows that S_{ROS} is approximately exponentially distributed when N is large and $\mu < N\lambda$. In this case, the number of customers Y at the repair facility varies relatively little over time, see, e.g. Remark 4.2, so that Y will be relatively close to its average value: $N(1 - \mu/\Lambda)$, where $\Lambda := N\lambda$. Ignoring the variability in Y, S_{ROS} is the sum of a random number, L, of $\exp(\mu)$ distributed service times, where L is geometrically distributed with parameter $1/((1 - \mu/\Lambda)N)$ (the tagged customer has a chance $1/j$ to be the next one served, if there are $j - 1$ other customers

Table 4.1. Mean $\mathbb{E}S_{ROS}$ and standard deviation σ_{ROS} of the sojourn times in the ROS model, for number of stations N, service rate μ, and total traffic intensity $\Lambda = N\lambda = 1$

μ	$N = 100$		$N = 200$		$N = 1,000$	
	$\mathbb{E}S_{ROS}$	σ_{ROS}	$\mathbb{E}S_{ROS}$	σ_{ROS}	$\mathbb{E}S_{ROS}$	σ_{ROS}
0.5	100	99	200	199	1,000	999
1.0	8.2	10.4	11.5	15.2	25.2	34.7
2.0	0.97	1.10	0.99	1.13	1.00	1.15

present). It is well known that the sum of a geometrically distributed number of independent, exponentially distributed random variables is again exponentially distributed, so that both mean and standard deviation of S_{ROS} can be approximated by $(1 - \mu/\Lambda)N/\mu$. In reality, Y will vary, so that this argument does not strictly apply: L is still geometrically distributed, but its parameter is a random variable rather than a constant equal to $1/(1-\mu/\Lambda)N$. Still, the entries in Table 4.1 with $\mu = 0.5$ show that the numerical predictions from this argument are excellent. In this case $1 - \mu/\Lambda = 0.5$, and both mean and standard deviation as numerically computed virtually coincide with the prediction $0.5N/\mu$. In fact, the numerical entries suggest that $\sigma_{ROS} = \mathbb{E}S_{ROS} - 1$. This can be made plausible by a more detailed version of the foregoing argument: By the arrival theorem the number of stations in the queue at the instants of departures is, on average, $(N-1)(1-\mu/\Lambda)$. Hence, the randomisation, which dominates the variance term, involves only $(N-1)(1-\mu/\Lambda)$ stations, on average, rather than $N(1-\mu/\Lambda)$.

REMARK 4.4 Let us briefly consider the other extreme case: $\mu/N\lambda \gg 1$. It is easily seen, and well known, that now $\mathbb{P}(X = N) \approx 1$. The repair facility now behaves like an open $M/M/1$ queue with arrival rate $\Lambda = N\lambda$ and service rate μ. Hence, $\mathbb{E}S = \mathbb{E}S_{ROS} \approx \frac{1}{\mu-\Lambda}$. In the standard $M/M/1$ queue with FCFS, the sojourn time is exponentially distributed, so $\sigma_{FCFS} = \mathbb{E}S$. In the $M/M/1$ queue with ROS, it follows from Cohen [35], p. 443, that the standard deviation of the sojourn time is inflated with a factor f as compared to the standard deviation of the standard $M/M/1$ queue with FCFS, where

$$f = \left(1 + \frac{2(\Lambda/\mu)^2}{2 - \Lambda/\mu}\right)^{\frac{1}{2}}.$$

The entries in Table 4.1 with $\mu = 2$ are relevant to this case. We find that $f = 1.15$, revealing a rather close agreement although $\frac{\mu}{N\lambda}$ only equals 2. Furthermore, note that $f \to 1$ for $\frac{\mu}{N\lambda} \gg 1$, which again yields a coefficient of variation of S_{ROS} that approaches 1.

REMARK 4.5 We finally consider the intermediate case $\rho = N - c\sqrt{N}$ as $N \to \infty$. This case has already been studied by Vaulot [168], see also Whitt [171], for the Erlang loss model with N servers and offered traffic ρ, a model that is equivalent with the repairman model. Vaulot proved that

$$\sqrt{N}B_N(\rho) = \sqrt{N}B_N(N - c\sqrt{N}) \sim \frac{\phi(c)}{\Phi(c)}, \quad c \in \mathbb{R}, \quad N \to \infty, \quad (4.25)$$

with

$$\phi(c) = \frac{1}{\sqrt{2\pi}}e^{-c^2/2} \quad \text{and} \quad \Phi(c) = \frac{1}{\sqrt{2\pi}}\int_{-\infty}^{c} e^{-x^2/2}dx$$

the standard normal density and standard normal distribution function, respectively. Substitution of (4.25) into (4.4) yields, for $c = 0$, the approximations

$$\mathbb{E}(X) \approx N - \sqrt{\frac{2N}{\pi}} \quad \text{and} \quad \mathbb{E}(Y) \approx \sqrt{\frac{2N}{\pi}}, \tag{4.26}$$

and hence, using Little's formula $\mathbb{E}Y = \Lambda_R \mathbb{E}S$, with $\Lambda_R = \lambda \mathbb{E}X$ the input rate into the repair facility:

$$\mathbb{E}(S) \approx \frac{1}{\Lambda} \sqrt{\frac{2N}{\pi}}. \tag{4.27}$$

The entries in Table 4.1 with $\mu = 1$ ($\rho = N$, $c = 0$) are relevant for this case.

4.5 GROS Discipline

In this section, we consider the model with the GROS discipline, as illustrated in Fig. 4.2 and described in Sect. 4.1. Again, we let Y denote the number of machines in the total waiting area (i.e. ante room plus waiting queue). Obviously, the distribution of Y equals that of the number of machines in the repair queue in the standard model described in Sect. 4.3, and is given by (4.2).

We will now consider S_{GROS}: the sojourn time until repair of an arbitrary (tagged) machine for the model with GROS. Observe that this sojourn time consists of two components:

$$S_{GROS} = \sum_{i=1}^{Y_{11}} B_{1i} + \sum_{i=1}^{Y_{21}} B_{2i}. \tag{4.28}$$

Here, the random variables B_{1i} and B_{2i} are independent, exponentially distributed service times with parameter μ. The random variable Y_{11} denotes the number of machines in the waiting queue (including the one in repair) at the instant that the tagged machine breaks down. The random variable Y_{21} equals the random position allocated to the tagged machine in the waiting queue at the instant it is moved from the ante room to the waiting queue.

This model is not a product-form network, so that an exact analysis of the sojourn time is considerably more difficult than the analysis for the models considered earlier. Still, a numerical analysis is possible. We first have to find the probability that the tagged machine finds k machines in the ante room and n machines in the waiting queue at the instant it breaks down, where $0 \leq k + n \leq N - 1$. Furthermore, a set of equations similar to (4.15) has to be solved, now for the functions $\phi_{k,n}(\omega)$, representing the LST of the sojourn time of a machine, given that it finds k machines in the ante room and n machines in the waiting queue at the instant that it breaks down.

A particularly simple evaluation of the first moments can be obtained, if one makes the following two approximating assumptions:

- The two components of S_{GROS} in (4.28) are uncorrelated.

- The random variables Y_{11} and Y_{21} are uniformly distributed on $\{1, 2, \ldots, Y_1\}$, with the random variable Y_1 as defined in Sect. 4.3.

In case $N\lambda > \mu$ and N is large, they appear to be good approximations. This can be motivated via Remarks 4.2 and 4.3: The two assumptions would be valid if there were a fixed fraction of the machines in the repair queue. This is not true, but the remarks suggest that this is a reasonable approximation as the variability in the number of machines in the repair queue is relatively small.

Under these assumptions, it is now straightforward to show that

$$\mathbb{E}(Y_{11}) \approx \frac{1}{2}\mathbb{E}(1 + Y_1) = \frac{1}{2}(N - \mu/\lambda),$$

and

$$\mathbb{E}(Y_{11}^2) \approx \frac{1}{3}\mathbb{E}(Y_1^2) + \frac{1}{2}\mathbb{E}(Y_1)$$

$$\approx \frac{1}{3}\left((N - 1 - \mu/\lambda)^2 + \mu/\lambda\right) + \frac{1}{2}(N - 1 - \mu/\lambda).$$

For the variance we find that

$$\mathrm{var}(Y_{11}) \approx \frac{1}{12}(N - \mu/\lambda)^2 - \frac{1}{6}N + \frac{1}{2}\mu/\lambda.$$

Using these approximations, we can now evaluate the moments of the sojourn time. The expected value is as in (4.10), and for the variance we obtain

$$\mathrm{var}(S_{GROS}) \approx 2\,\mathrm{var}\left(\sum_{i=1}^{Y_{11}} B_{1i}\right) = \frac{2}{\mu^2}(\mathbb{E}(Y_{11}) + \mathrm{var}(Y_{11}))$$

$$\approx \frac{1}{\mu^2}\left(\frac{(N - \mu/\lambda)^2}{6} + \frac{2N}{3}\right). \tag{4.29}$$

Now the variance of the sojourn time is of intermediate magnitude for large N. It is much larger than in the FCFS case, but much smaller than in the ROS case.

4.6 GPROS Discipline

The sojourn time until repair of an arbitrary job for the model with GPROS, denoted by S_{GPROS}, allows for a heuristic approximation that is similar to the one given for GROS. For this, observe that this sojourn time consists of three components:

$$S_{GPROS} = \sum_{i=1}^{Z_{11}} B_{1i} + \sum_{j=1}^{G-1}\sum_{i=1}^{Z_j} A_{ij} + \sum_{i=1}^{Z_{21}} B_{2i}. \tag{4.30}$$

Here, the random variables B_{1i}, B_{2i} and A_{ij} are independent, exponentially distributed service times with parameter μ. The random variable Z_{11} represents the, remaining, number of machines of the batch currently in the repair queue (including the one in service) at the instant that the tagged machine breaks down. The random variable Z_{21} represents the random service position of the tagged machine within its batch. Finally, the random variables $Z_j, j = 1, \ldots, G - 1$, denote the sizes of the batches that are served between the batch in service at the instant of the arrival of the tagged machine and the batch to which the tagged machine belongs.

We now make even stronger approximating assumptions than in the previous section:

- The random variable G always takes the value g.

- The three components of S_{GPROS} in (4.30) are uncorrelated.

- The random variables Z_{11} and Z_{21} are uniformly distributed on $1, \ldots, Y_1/g$, where the random variable Y_1 is as defined in Sect. 4.3.

- The random variables $Z_j, j = 1, \ldots, g-1$, are distributed as Y_1/g, suitably rounded.

Again, neither of these assumptions is strictly valid. However, for $N\lambda > \mu$, N large, and g small, they are reasonable as the variability of the number of machines in the repair queue is relatively small, see Remarks 4.2 and 4.3 and the discussion following the approximating assumptions for the GROS model.

As in Sect. 4.5, it immediately follows that

$$\mathbb{E}(Z_{11}) \approx \frac{1}{2}\frac{1}{g}\mathbb{E}(1 + Y_1) = \frac{1}{2}\frac{1}{g}(N - \mu/\lambda),$$

and

$$\mathbb{E}(Z_{11}^2) \approx \frac{1}{g^2}\frac{1}{3}\mathbb{E}(Y_1^2) + \frac{1}{2}\frac{1}{g}\mathbb{E}(Y_1)$$

$$\approx \frac{1}{3}\frac{1}{g^2}\left((N - 1 - \mu/\lambda)^2 + \mu/\lambda\right) + \frac{1}{2}\frac{1}{g}(N - 1 - \mu/\lambda).$$

For the variance we find that

$$\operatorname{var}(Z_{11}) \approx \frac{1}{12}\frac{1}{g^2}(N - \mu/\lambda)^2 - \frac{2}{3}\frac{1}{g^2}N + \frac{1}{g^2}\mu/\lambda + \frac{1}{2}\frac{1}{g}(N - \mu/\lambda).$$

For the moments of the Z_j we obtain

$$\mathbb{E}(Z_1) \approx \frac{1}{g}\mathbb{E}(Y_1) = \frac{1}{g}(N - 1 - \mu/\lambda)$$

and
$$\mathbb{E}(Z_1^2) \approx \frac{1}{g^2}\mathbb{E}(Y_1^2) \approx \frac{1}{g^2}\left((N-1-\mu/\lambda)^2 + \mu/\lambda\right).$$

For the variance we find
$$\text{var}(Z_1) \approx \frac{\mu}{\lambda g^2}.$$

Using these approximations, we can now approximate the moments of the sojourn time.

The expectation of the sojourn time is as before and given in (4.10). To approximate the variance, we first use the assumption that G can be replaced by g:

$$\text{var}(S_{GPROS}) \approx \text{var}\left(\sum_{i=1}^{Z_{11}} B_{1i} + \sum_{j=1}^{g-1}\sum_{i=1}^{Z_j} A_{ij} + \sum_{i=1}^{Z_{21}} B_{2i}\right)$$

$$= 2\text{var}\left(\sum_{i=1}^{Z_{11}} B_{1i}\right) + (g-1)\text{var}\left(\sum_{i=1}^{Z_1} A_{i1}\right)$$

$$= \frac{2}{\mu^2}\left(\mathbb{E}(Z_{11}) + \text{var}(Z_{11})\right) + \frac{g-1}{\mu^2}\left(\mathbb{E}(Z_1) + \text{var}(Z_1)\right).$$

$$(4.31)$$

One can now substitute the expressions for the moments of Z_1 and Z_{11} to obtain an explicit expression.

Inspecting the expression for $\text{var}(S_{GPROS})$, we see that the variance is characterised by two regimes. Firstly, for g fixed and $N \to \infty$, the quadratic term in $\text{var}(Z_{11})$ dominates, so that

$$\text{var}(S_{GPROS}) \approx \frac{2}{\mu^2}\text{var}(Z_{11})$$

$$\approx \frac{1}{6g^2\mu^2}(N-\mu/\lambda)^2.$$

In this case, the GPROS discipline approximately reduces the variance of the sojourn time by a factor g^2 as compared to GROS. Secondly, assume that $g = c(N-\mu/\lambda)$ for some constant $c > 0$, as suggested by the results on skipping levels in Sect. 3.3.3. Then we find that

$$\text{var}(S_{GPROS}) \approx \frac{g-1}{\mu^2}\mathbb{E}(Z_1) \approx \frac{g-1}{g\mu^2}(N-\mu/\lambda) \approx (N-\mu/\lambda)/\mu^2$$

for $N \to \infty$. Hence, in this case, the variance of the sojourn time is approximately equal to the variance of the sojourn time for FCFS.

4.7 Numerical Results

We now turn to a comparison of the access delay due to contention resolution and the sojourn time in the variants of the repairman model. In this comparison, we will confine ourselves to the first two moments of the various random variables. We consider first moments in Sect. 4.7.1 and standard deviations in Sect. 4.7.2.

The procedures for contention resolution were described in Sect. 4.2, and the involved access delay is the delay experienced by stations that use contention trees for reservation. More formally, it is defined as the number of tree slots elapsed from the instant a station becomes active until the instant its request is successfully transmitted. As already indicated, there are no closed-form expressions for the moments of the access delay, and so the results are obtained by simulation. In these simulations, the stations execute one of the procedures outlined in Sect. 4.2: They become active after an exponentially distributed inactive period with parameter λ. Then they enter the contention resolution at the earliest possible instant, as defined by the channel access protocol. Thus, we use a source model in which each of the N stations generates packets according to a Poisson process with rate λ, independently of the other stations.

The average delays obtained are denoted $\widehat{\mathbb{E}S_F}$ and $\widehat{\mathbb{E}S_B}$, for the 'free' and 'blocked' channel access protocol, respectively. Likewise, the estimated standard deviations are denoted by $\widehat{\sigma_F}$ and $\widehat{\sigma_B}$. The 'hat' serves as a reminder that the moments are obtained from simulation. The figures for the blocked trees are obtained by simulating 5,000 trees at the indicated settings. For the free trees, the channel was simulated for 500,000 tree nodes. The figures presented are sample averages after an initialisation period of 10%. By analysing the data as five separate batches, we obtained an indication of the standard deviation of the estimated figures. This analysis shows that all presented estimates have a standard deviation of less than 1% of the estimate itself, which is sufficient for our purposes.

The moments of the sojourn time of the various repairman models have been obtained in Sects. 4.3–4.6. In utilising the results from these sections, we will use $\mu = \log(3)$ for the service rate. The motivation behind this value is in Theorem 3.1 from Chap. 3 which gives the average length of a contention tree with n contenders. Janssen and de Jong [86], (26)–(27), show that the average length is well approximated by $n/\log(3)$. Hence, the rate at which the contenders are served can be approximated by $\log(3)$.

When approximating skipped level trees by means of the repairman model with GPROS, we no longer have such a simple approximation for the service rate. In this case, the arguments in Sect. 3.3.3 suggest that we may proceed as follows. Firstly, we obtain a crude approximation to the rate using (3.51) in Chap. 3:

$$\mu^{(opt)} = m \left(m \sum_{l=-1}^{\infty} h(\xi m^{-l}) + \frac{1}{m\xi} \right)^{-1}. \qquad (4.32)$$

Compared to (3.51), we have substituted $s = k - 1$ as motivated by the discussion below (3.51), and have multiplied by m to account for the fact that there are m minislots in one slot. Secondly, we evaluate (4.32) for $\xi = 2$ and $m = 3$, to obtain $\mu^{(opt)} \approx 1.2$. Thirdly, we compute the expected number of contenders in each contention tree, $\mathbb{E}Y$, according to Remark 4.2. Finally, we compute the optimal number of levels to skip as

$$s^{(opt)} = \left\lfloor \overset{3}{\log} \left(\mathbb{E}Y \right) \right\rfloor. \qquad (4.33)$$

Note that the initial number of subgroups equals $g = m^{s+1}$ if the first s levels are skipped.

4.7.1   First Moments

The average access delays for the tree models and the expected sojourn time for the repairman model are given in Table 4.2. There is only one entry in the table corresponding to the expected sojourn time, as it is the same for all variants of the repairman model considered. In the table, we have varied the number of stations, N, and the total traffic intensity $\Lambda := N\lambda$. The primary purpose of this table is to compare average access delay with expected sojourn time. The intensities are chosen so that Λ is well above μ, which corresponds to the setting for which the finite-population approximations are valid.

We can draw various conclusions. Firstly, and most importantly, we observe that the expected sojourn time in the repairman model provides an excellent approximation to the average access delay for reservation with contention trees. The agreement with blocked access is almost perfect; the agreement with the results for free access is somewhat less. The former result is closely related to a result in Denteneer and Pronk [55] on the average number of contenders in a contention tree.

Table 4.2. Average request delay with free trees, $\widehat{\mathbb{E}S_F}$, and with blocked trees, $\widehat{\mathbb{E}S_B}$, both obtained via simulation, and expected sojourn time for the repairman model, $\mathbb{E}S$, for number of stations N, and total traffic intensity Λ

	$N = 100$			$N = 200$			$N = 1{,}000$		
Λ	$\widehat{\mathbb{E}S_F}$	$\widehat{\mathbb{E}S_B}$	$\mathbb{E}S$	$\widehat{\mathbb{E}S_F}$	$\widehat{\mathbb{E}S_B}$	$\mathbb{E}S$	$\widehat{\mathbb{E}S_F}$	$\widehat{\mathbb{E}S_B}$	$\mathbb{E}S$
2.5	42.9	50.1	51.0	85.6	100.9	102.0	430.2	509.2	510.0
5.0	62.9	70.4	71.0	125.7	141.3	142.0	628.6	710.3	710.0
10.0	72.7	80.5	81.0	145.6	161.4	162.0	729.2	809.7	810.0
16.5	76.7	84.5	84.9	153.4	169.4	169.9	769.2	849.2	850.0

Table 4.3. Average request delay for skip level trees, $\widehat{\mathbb{E}S}_S$, with g initial slots, and expected sojourn time for the repairman model, $\mathbb{E}\tilde{S} := \mathbb{E}S_{GPROS}$, with the GPROS discipline, for number of stations N, and total traffic intensity Λ

Λ	$N = 100$			$N = 200$			$N = 1,000$		
	g	$\widehat{\mathbb{E}S}_S$	$\mathbb{E}\tilde{S}$	g	$\widehat{\mathbb{E}S}_S$	$\mathbb{E}\tilde{S}$	g	$\widehat{\mathbb{E}S}_S$	$\mathbb{E}\tilde{S}$
2.5	27	42.6	43.3	81	81.5	86.7	729	447	433
5.0	27	65.5	63.3	81	124.8	126.7	729	609	633
10.0	81	70.6	73.3	81	147.2	146.7	729	707	733
16.5	81	74.1	77.3	81	155.5	154.5	729	746	772

Secondly, we see that free access is more efficient than blocked access in that the former has a smaller average access delay. This result parallels the result for the open model and the Poisson source model, as graphically illustrated in Fig. 16 of Mathys and Flajolet [120]. The considered variants of the repairman model all lead to the same expected sojourn time and are apparently not sufficiently detailed to capture the differences between the first moments of the blocked and the free access protocols. The difference arises as free and blocked trees have a different efficiency, see Sect. 2.1. Indeed, the entries from Table 4.2 suggest that free trees are approximately 10% more efficient than blocked trees, in agreement with Table 2.1. Finally, we observe that all quantities displayed in Table 4.2 approximately exhibit a linear dependence on the number of stations (for the cases with $N \gg \mu/\lambda$).

We now consider the approximation to the sojourn time in scheduled contention trees using the repairman model with the GPROS discipline. Some values for the expected sojourn time are given in Table 4.3. In all, the approximations are useful.

4.7.2 Standard Deviations

We next turn to a numerical comparison of the standard deviations in the various models. These are given in Table 4.4, again for various values of N and Λ.

Several conclusions can be drawn. Firstly, we observe that the standard deviation in either tree model changes with the traffic intensity and grows approximately linearly with the number of stations. Neither of these properties is captured by the basic repairman model. In this basic model, the standard deviation of the sojourn time is independent of the traffic intensity and grows only as the square root of the number of stations.

Secondly, the standard deviation of the access delay in the blocked tree model agrees well with the corresponding figure for the GROS repairman model. The difference between the two standard deviations is approximately 15%.

Table 4.4. Standard deviations of the request delay with free trees, $\widehat{\sigma_F}$, with blocked tree, $\widehat{\sigma_B}$, and standard deviations for the basic repairman model, σ, the ROS repairman model, σ_{ROS}, and the GROS repairman model, σ_{GROS}, for number of stations N, and total traffic intensity Λ

		Tree		Repair		
N	Λ	$\widehat{\sigma_F}$	$\widehat{\sigma_B}$	σ	σ_{ROS}	σ_{GROS}
100	2.5	46.3	19.3	9.1	50.45	22.1
100	5.0	67.9	26.5	9.1	70.18	29.9
100	10.0	78.5	30.1	9.1	80.13	33.9
100	16.5	82.8	31.4	9.1	84.06	35.5
200	2.5	92.5	37.5	12.9	101.46	43.0
200	5.0	136.0	52.5	12.9	141.20	58.9
200	10.0	157.4	59.9	12.9	161.15	67.0
200	16.5	165.4	62.4	12.9	169.02	70.1
1,000	2.5	464.3	184.9	28.8	509.64	209.6
1,000	5.0	677.9	260.9	28.8	709.39	290.9
1,000	10.0	789.8	298.4	28.8	809.34	331.6
1,000	16.5	834.6	310.6	28.8	848.73	347.7

The results for the GROS model capture both the dependence on the traffic intensity and the dependence on the number of machines that is observed in the tree simulations. Similarly, the standard deviation of the access delay in the free tree model matches the corresponding figure for the repairman model with the ROS service discipline quite well.

Looking more closely at the results, we see that the standard deviations obtained for the GROS repairman model are always larger than those obtained in the blocked tree simulations. Here, a fundamental limitation of the repairman model as an approximation shows up. The batch nature of the contention trees implies that it takes some initial time before the first successful request is transmitted. Theorem 3.3 states that, after this initial period, successful transmissions occur fairly uniformly over the length of the trees. Thus, the variability of the waiting period is somewhat reduced as compared to the proposed model in which the successful transmissions occur uniformly over the full length of the tree. This also suggests that there is an even more appropriate extension of the basic repairman model, i.e. one in which the transfer from the ante room to the queue takes some time and in which the server operates at a slightly higher speed. Alternatively, the current approach is more appropriate for skip level trees as described in Sect. 3.3.3.

Thirdly, the standard deviations for the free access protocol far exceed those for the blocked access protocol. This result has no counterpart in the open model. In fact, Fig. 17 in Mathys and Flajolet [120] shows that the standard deviation of the delay for free access protocol is below the corresponding value with blocked access for most traffic intensities. However, for large traffic

Table 4.5. Standard deviation of request delay for skip level trees, $\widehat{\sigma_S}$, with g initial slots, and standard deviation of the sojourn time for the GPROS repairman model, σ_{GPROS}, for number of stations N, and total traffic intensity Λ

	$N = 100$			$N = 200$			$N = 1{,}000$		
Λ	g	$\widehat{\sigma_S}$	σ_{GPROS}	g	$\widehat{\sigma_S}$	σ_{GPROS}	g	$\widehat{\sigma_S}$	σ_{GPROS}
2.5	27	7.1	6.0	81	9.4	8.5	729	18.2	19.0
5.0	27	6.4	7.3	81	8.9	10.3	729	20.5	23.0
10.0	81	6.2	7.8	81	8.9	11.1	729	20.3	24.8
16.5	81	6.3	8.0	81	8.5	11.4	729	19.2	25.4

intensities just below the stability bound the order reverses, and blocked access then results in smaller standard deviations. Of course, our simulations are carried out at a totally different set of traffic intensities, that exceed the stability bound for the open system.

Moreover, the approximations exceed the values for the standard deviations obtained in simulation. This observation parallels a similar finding for the first moments when dealing with free trees, see Table 4.2. The correspondence between the approximations and the values obtained in simulation can be further improved by taking into account that the free trees operate at a slightly higher rate than the blocked trees. One possibility is to insert the rate of free trees, obtained from Table 2.1, in the expressions for the standard deviations.

We now consider the approximation to the sojourn time in scheduled contention trees using the repairman model with the GPROS discipline. Some values are given in Table 4.5. The approximation of the standard deviation of the sojourn time, considered as a function of N, is rather good. The approximation, considered as a function of Λ, however, leaves room for improvement: The approximations are linearly increasing in Λ, whereas the values obtained by simulation are relatively insensitive to Λ.

4.8 Conclusion

In this chapter, we started from the assumption that the sojourn time in the repair facility of the repairman model is related to the access delay experienced when using contention trees to transmit requests. To substantiate this premise, we obtained expressions for the moments of this sojourn time and compared these to the corresponding moments of the access delay obtained through simulation. These numerical experiments showed that the expected sojourn time in the repair stage shows a perfect match with the average access delay for both variants of the tree procedure. It was also shown that the variances of the access delay for the various tree protocols are well approximated by the variances of the sojourn times of the appropriate repairman model: ROS for free access, GROS for blocked access, and GPROS for scheduled blocked access.

In the introduction of this chapter it was pointed out that the variance of the access delay in the first stage is needed in understanding the average total packet delay in the two stages of a reservation procedure. The analysis of the total delay will be carried out in Chap. 11. Moreover, the analysis of the models with the GROS and GPROS disciplines was heuristic in nature. Further research is needed to make these approximations more precise. We take a first step in this direction in Chap. 5.

An interesting topic not considered in this book, is to investigate the extent to which the repairman model can be used to model other random access algorithms, in which a finite-population effect is expected to be important. We refer to Winands et al. [174] for such a study in case of ALOHA with exponential back-off, used in the context of wireless in-home networks.

A final comment pertains to scheduled access, described in Sect. 4.2.1, and introduced specifically to decrease the request variability. However, this can alternatively be achieved by allowing access only for those stations that have at least a certain minimum number of packets to transmit. The comparison of this method with the scheduled access is another topic for further research.

Chapter 5

THE REPAIRMAN MODEL WITH GROS

The repairman model is used in Chap. 4 to approximate the request delay in a reservation procedure. In particular, the sojourn time in the repairman model with the Gated Random Order of Service (GROS) discipline is suggested as an approximation for the request delay, when blocked contention trees are used in a dynamic environment. Section 4.5 gives a heuristic approximation to the variance of the sojourn time under GROS. That approximation would be valid under overload conditions when the number of stations is large.

In this chapter, we undertake a more detailed analysis of the queue-size processes in this closed queueing system with the GROS discipline via fluid approximations. We use a continuous-mapping approach to carry out the asymptotic analysis. Additionally, we use the fluid approximation to the queue-size processes to study the empirical measure defined by the sojourn times of the jobs that enter between two gate openings. Our analysis shows that this empirical measure converges to the convolution of two uniform distributions, which confirms the heuristic approximation from the previous chapter.

5.1 Introduction

The repairman model with the Gated Random Order of Service (GROS) discipline was introduced in Chap. 4. The system is depicted in Fig. 4.2, and its operation can be described as follows. The waiting area consists of an ante room and a queue with a gate in between. Arriving jobs are placed in this ante room before the queue. As soon as the single server becomes idle, the gate opens and all jobs in the ante room are placed in the queue. The gate then closes, and the jobs in the queue are served in random order. If there are no such jobs, i.e. if the ante room is empty, the gate remains open until the first arrival. This job is immediately transferred to the queue, and the gate closes.

In Chap. 4, it was shown that the sojourn time in the repairman model with the GROS discipline constitutes an appropriate approximation to the request delay, when using blocked contention trees in a dynamic environment. Moreover, in (4.29) of Chap. 4 a heuristically motivated approximation to the variance of this sojourn time is given. This approximation would be valid under overload conditions when the number of stations is large.

In this chapter, we undertake a more detailed analysis of the queue-size processes in this closed queueing system with the GROS discipline via fluid approximations. The analysis exploits the similarity of the queueing system with the GROS discipline and the same closed queueing system with the FCFS discipline. In particular, it is readily verified that the arrival process to the waiting area associated with the single server and the departure process from the single server are stochastically the same for either service discipline. Moreover, fluid approximations for these arrival and departure processes for the system with the FCFS discipline have been well investigated, see, e.g. Iglehart and Whitt [82] and Krichagina and Puhalskii [103].

The following is then a natural approach to analyse the queue-size processes. Firstly, define a map that transforms the arrival and departure processes into the queue-size processes, and investigate the continuity of this map. Secondly, consider the limits of the arrival and departure processes under fluid scaling. This latter limit can be obtained from the references cited, or, under more restrictive conditions, from general theory given in Mandelbaum et al. [117]. Thirdly, obtain the scaling limits of the relevant queue-size processes by combining the scaling limits in the base system with the properties of the map. In this approach, we have followed Whitt [172], Chap. 3. In the conclusion, we comment on an extension of this approach which additionally uses the directional derivative of the map and the diffusion limits of the arrival and departure processes, see [172], Internet Supplement Chap. 9. Further examples of this approach can be found in, e.g. Mandelbaum and Massey [116] and Mandelbaum and Pats [118].

Next, we show that the fluid approximation to the queue-size processes can be used to approximate the sojourn times of the machines in repair. More in particular do we study the empirical measure defined by the sojourn times of the jobs that enter between two gate openings. Here, our analysis shows that this empirical measure converges to the convolution of two uniform distributions.

In Sect. 5.2, we first give a precise description of the model under consideration. Then, in Sect. 5.3, we review the approach and the main results and give an outline of the rest of this chapter.

5.2 Model Description

We consider a closed queueing system with N jobs, two service stations, and an ante room in between, as illustrated in Fig. 4.2. The first station is an infinite server queue and the second station is a single server queue. Jobs move cyclically through the three nodes in the system as follows: After having been served at the infinite server queue, a job moves to the ante room, where it waits behind a gate in front of the single server queue. At the instant the single server becomes idle, the gate opens and all jobs in the ante room are placed in the single server queue. If there is no job waiting in the ante room, the gate remains open until a first job arrives. This job is immediately taken into service at the single server. The gate then closes. Jobs that are moved from the ante room are placed in the single server queue in random order, and are served in this order.

This system is similar to the system considered in Krichagina and Puhalskii [103] except for the gate and the ante room. We now give a precise description of the system, following [103]. This description defines a sequence of models indexed by N, the number of jobs in the system. In the scaling analysis to follow, we will let N tend to infinity, and the investigation will be relevant to large systems.

Assume that there are initially N_0 jobs in the ante room, that there are no jobs in the single server queue, so that there are $N - N_0$ jobs in the infinite server queue. Next, assume there is a complete probability space $(\Omega, \mathcal{F}, P)$ and independent i.i.d. sequences $\{\eta_i, i \geq 1\}$, $\{\tilde{\eta}_i, i = 1, \ldots, N - N_0\}$, $\{\xi_i, i \geq 1\}$, and $\{u_i, i \geq 1\}$ of nonnegative random variables defined on $(\Omega, \mathcal{F}, P)$. The random variables $\{\tilde{\eta}_i, i = 1, \ldots, N - N_0\}$ denote the residual service times of the jobs initially present in the infinite server queue. The random variables $\{\eta_i, i \geq 1\}$ represent the consecutive service times associated with jobs entering the queue at the infinite server and have a common distribution with cumulative distribution function $F(x)$. We assume that $E(\eta_1) = 1/\lambda$.

The variables $\{\xi_i, i \geq 1\}$ denote the consecutive service requirements of the jobs served at the single server. We assume that $E(\xi_1) = 1$ and that $\text{var}(\xi_1) < \infty$. At this single server, the jobs are served with rate $N\mu$. Note that this involves a scaling of the rate at which the single server operates. We do so as N tends to infinity in the proposed scaling limits. This will result in an arrival intensity at the single server queue of rate proportional to N. In order to obtain relevant limits, this arrival intensity must be balanced by a similar increase of the rate at which the single server operates.

The sequence of random variables $\{u_i, i \geq 1\}$ is independent of the other sequences and consists of independent random variables, each uniformly distributed on $[0, 1]$. They will be used to model the randomness in the service discipline.

We use $V^N(t)$ to denote the number of jobs being served by the infinite server at time t. Moreover, we use $W^N(t)$ for the number of jobs in the ante room at time t. The number of jobs at the single server at time t is denoted by $Q^N(t)$, and this number includes both the job in service and the jobs in the queue in front of this single server. Finally, let $Y^N(t) = W^N(t) + Q^N(t)$ be the number of jobs that are either in the ante room or at the single server, so that $V^N(t) = N - Y^N(t)$.

The queue-size processes $V^N(\cdot)$, $Y^N(\cdot)$, $W^N(\cdot)$, and $Q^N(\cdot)$ can be described in terms of the primitive random variables as follows. Let $S(\cdot) = \{S(t), t \geq 0\}$ denote the renewal process generated by the sequence $\{\xi_i\}$:

$$S(t) = \max(k : \xi_1 + \cdots + \xi_k \leq t). \tag{5.1}$$

Let $D^N(\cdot)$ denote the departure process from the single server queue, i.e. $D^N(t)$ is the number of jobs that have been served by the single server by time t:

$$D^N(t) = S\left(N\mu \int_0^t \mathbf{1}(V^N(s) < N)ds\right). \tag{5.2}$$

The departure process for the single server queue is also the arrival process at the infinite server queue. Hence, if $\{\alpha_i^N, i \geq 1\}$ is the sequence of arrival times of jobs at the infinite server, then

$$\alpha_i^N = \inf(t \geq 0 : D^N(t) \geq i). \tag{5.3}$$

For each $t \geq 0$, the arrival process, $A^N(\cdot)$, at the ante room is given by

$$A^N(t) = \sum_{i=1}^{D^N(t)} \mathbf{1}(\alpha_i^N + \eta_i \leq t) + \sum_{j=1}^{N-N_0} \mathbf{1}(\tilde{\eta}_j \leq t). \tag{5.4}$$

Clearly, $A^N(\cdot)$ is also the departure process from the infinite server queue, i.e. $A^N(t)$ is equal to the number of jobs that have completed service at the infinite server by time t. Using $A^N(\cdot)$, we can define the arrival time at the ante room of the jth job to leave the infinite server:

$$\sigma_j^N = \inf(t \geq 0 : A^N(t) \geq j). \tag{5.5}$$

The total number of jobs in the ante room and the single server queue equals

$$Y^N(t) = N_0 + A^N(t) - D^N(t). \tag{5.6}$$

Note that the definition of this system of equations is circular as $D^N(\cdot)$ depends on $V^N(\cdot)$ and $V^N(\cdot)$ depends, via $Y^N(\cdot)$, on $D^N(\cdot)$. However, it is easy to see that there is exactly one solution to this system of equations, as we can distinguish between busy periods of the single server and idle periods. During busy periods $V^N(t) < N$, so that the circularity of the definition vanishes. The idle periods are easily seen to be uniquely defined.

REMARK 5.1 The model definition so far has closely followed [103], where a gate-less system is considered. Apart from the gate, there are some further changes, as compared to [103]. Firstly, we have not assumed that the infinite server queue is initially empty. Secondly, we assume that the rate of the single server is fixed and equal to $N\mu$, whereas in [103] the rate depends on the size of the queue in front of the single server.

We now proceed to introduce additional notation, due to the introduction of the ante room and the gate. These are the instants of the gate openings, $\tau_j^N, j = 0, 1, 2, \ldots$, and the number of jobs, $N_j, j = 0, 1, 2, \ldots$, present in the ante room immediately before the instants of these gate openings. First, define $\tau_0^N = 0$ and assume that $N_0 > 0$ is known. Next, define

$$\tau_1^N = \max\left(\alpha_{N_0}^N, \sigma_1^N\right), \tag{5.7}$$

and for $j = 1, 2, \ldots$

$$\tau_{j+1}^N = \max\left(\alpha_{N_0+A^N(\tau_j^N)}^N, \sigma_{A^N(\tau_j^N)+1}^N\right), \tag{5.8}$$

and

$$N_j = A^N(\tau_j^N) - A^N(\tau_{j-1}^N). \tag{5.9}$$

By construction, we have $N_j \geq 1$, for all j. The group of jobs that arrive during one gate period will be called a batch, and N_j is the batch size.

Using these definitions, we can express the number of jobs in the ante room at time t as follows:

$$W^N(t) = A^N(t) - \max_{\tau_i^N \leq t} A^N(\tau_i^N), \tag{5.10}$$

and the number of jobs at the single server

$$\begin{aligned} Q^N(t) &= Y^N(t) - W^N(t) \\ &= N_0 + \max_{\tau_i^N \leq t} A^N(\tau_i^N) - D^N(t). \end{aligned} \tag{5.11}$$

Next, we turn to sojourn times. For this, we must model the randomisation procedure that characterises GROS. This is done as follows. On its arrival at the ante room, a job is assigned the next value from the sequence $\{u_i, i \geq 1\}$. Thus the last job to arrive at the ante room before time t is the $A^N(t)$th arrival to the ante room and is assigned the value

$$u_{A^N(t)}. \tag{5.12}$$

These values are used to determine the order in which the jobs are served in the single server queue. More specifically, let i_1 and i_2 denote the arrival location of two jobs, i.e. the first job is the i_1th arrival to the ante room and the second job is the i_2th arrival to the ante room. Assume that both jobs arrive at the ante room during the same cycle, j, say, so that

$$\sum_{l=1}^{j-1} N_l + 1 \le i_1 < i_2 \le \sum_{l=1}^{j} N_l. \tag{5.13}$$

Then the job associated with the i_1th arrival is served before the job associated with the i_2th arrival if $u_{i_1} \le u_{i_2}$. Conversely, the job associated with i_2 is served first if $u_{i_1} > u_{i_2}$. It is clear that this mechanism induces the right order at the single server queue.

5.2.1   Notation and Conventions

We now introduce some notation that is used throughout the chapter. We use $\mathbf{D}$ for the space of functions from $[0, \infty)$ to $\mathbb{R}^k$ that are right-continuous and have left-hand limits, and $\mathbf{D}$ will be endowed with the product Skohorod J_1-topology. We use subscript symbols to indicate subspaces of $\mathbf{D}$: $\mathbf{D}_u$ denotes the subset of functions in $\mathbf{D}$ that are unbounded above, $\mathbf{D}_\uparrow$ denotes the subset of functions in $\mathbf{D}$ that are non-decreasing, and $\mathbf{D}_{\uparrow\uparrow}$ stands for the subset of functions in $\mathbf{D}$ that are strictly increasing. We use $\mathbf{C}$ to denote the subspace of $\mathbf{D}$ of continuous functions and similar subscript symbols are used to denote the subsets of $\mathbf{C}$. These are all Borel measurable subspaces of $\mathbf{D}$. The subspaces of $\mathbf{D}$ are endowed with the restrictions of the J_1-topology. For the spaces of continuous functions, this restriction coincides with the uniform topology.

The superscript T is used to indicate the restriction to a function space with compact domain, so that $\mathbf{D}^T$ denotes the space of functions from $[0, T]$ to $\mathbb{R}^k$ that are right-continuous and have left-hand limits.

Convergence with respect to the J_1-topology will be denoted by $\overset{J_1}{\rightarrow}$:

$$x^\epsilon \overset{J_1}{\rightarrow} x$$

for $\epsilon \to 0$. Consider the metric

$$d_{J_1}^T(x, y) = \inf_{\nu \in \mathcal{H}_T} \left(\|x \circ \nu - y\|_T \vee \|\nu - e\|_T \right),$$

with $a \vee b := \max(a, b)$ and $\mathcal{H}_T$ the set of homeomorphisms of $[0, T]$ onto $[0, T]$: Strictly increasing and continuous functions that map $[0, T]$ onto $[0, T]$. In order to prove J_1-convergence in $\mathbf{D}$ it is necessary to prove convergence with respect to the metric $d_{J_1}^T$ for every $T > 0$ except a countable set.

We use $\mathcal{H}$ to denote the space of continuous and strictly increasing functions. Finally, we use $\Rightarrow$ to denote convergence in distribution for sequences of random elements in $\mathbf{D}$, see Whitt [172], Sect. 3.2. The function e will be used to denote the identity, so that $e(t) = t$ for all t, and $\mathbf{1}$ is used for the indicator function.

5.3   Approach and Main Results

In (5.10) and (5.11) we have expressed the queue processes $W^N(\cdot)$ and $Q^N(\cdot)$ as a transformation of the arrival and departure processes $A^N(\cdot)$ and $D^N(\cdot)$. These arrival and departure processes are stochastically the same for the closed queueing system with the GROS discipline and the closed queueing system with the FCFS discipline, which has been studied in, e.g. Iglehart and Whitt [82] and Krichagina and Puhalskii [103]. In order to infer the scaling limits of the queue processes from the scaling limits of the arrival and departure processes, we set up a map, Γ, that maps the latter processes to the former processes.

To define Γ, consider function pairs $(d, a) \in \mathbf{D}^* := \mathbf{D}_u \times \mathbf{D}_\uparrow$ that satisfy

$$d(0) = 0, \quad a(0) = a_0,$$

for a real-valued quantity $a_0 > 0$. Next, define $\tau_0 := 0$ and recursively:

$$\tau_1(d, a) \quad := \quad \inf_s(d(s) \geq a(0)),$$
$$\tau_2(d, a) \quad := \quad \inf_s(d(s) \geq a(\tau_1)),$$

$$\vdots$$

$$\tau_i(d, a) \quad := \quad \inf_s(d(s) \geq a(\tau_{i-1})). \tag{5.14}$$

Finally, define τ^* as the limit point of this sequence,

$$\tau^* := \lim_{j \to \infty} \tau_j(d, a), \tag{5.15}$$

which may be finite or infinite. The first j such that the limit τ^* is attained will be denoted by j^*

$$j^* := \min(j : \tau_j = \tau^*), \tag{5.16}$$

and j^* can be finite or infinite. We note that the τ_i form a strictly increasing sequence until the limit τ^* is reached, i.e. $\tau_i > \tau_{i-1}$ for $i \leq j^*$.

Define $\gamma(t)$ to equal the value of the last transfer instant up to time t for (d, a):

$$\gamma(d, a)(t) := \max(\tau_i : \tau_i \leq t). \tag{5.17}$$

The transfer map Γ is then defined as

$$\Gamma \begin{pmatrix} d \\ a \end{pmatrix} = \begin{pmatrix} -d + a \circ \gamma \\ a - a \circ \gamma \end{pmatrix}, \tag{5.18}$$

with γ as in (5.17).

The connection with the GROS system as defined in Sect. 5.2 should be obvious: The functions (d, a) could form a departure process from the single server queue and an arrival process at the ante room, respectively, as defined in (5.2) and (5.4). These τ_j are then the transfer instants of a GROS system until the first special gate period, i.e. until the first cycle that commences with an empty ante room. After this first special gate period the processes stop. Using $(q, w) := \Gamma(d, a)$, it is then clear that q equals the number of jobs at the single server queue, cf. (5.11), up to the first special gate period. Similarly, w equals the number of jobs in the ante room, cf. (5.10), up to the first special gate period.

We shall prove, see Sect. 5.4.1, that Γ is a measurable mapping from $\mathbf{D}^*$ to $\mathbf{D}$ and, see Lemma 5.1, that Γ is continuous in the Skorohod J_1-topology on the set of continuous increasing functions that satisfy a rate constraint. These properties enable us to invoke the continuous-mapping theorem, see, e.g. Whitt [172], Theorem 3.4.3, and infer the fluid limits of $Q^N(\cdot), W^N(\cdot)$ from those of $D^N(\cdot), A^N(\cdot)$.

In particular, we consider a sequence of closed queueing systems with the GROS discipline indexed by N, as defined in Sect. 5.2. We assume that the service requirements at both the infinite server and the single server are exponential. In the conclusion to this chapter, we comment on less restrictive scenarios. The distributions of the service requirements $\{\eta_i\}$ and $\{\tilde{\eta}_i\}$ are exponential with rate λ. The service requirements at the single server $\{\xi_i\}$ are exponential with mean 1. The service rate equals $N\mu$.

Define $X^N := (D^N, N_0 + A^N)/N$ and assume that the system starts in steady state at the instant of a gate opening:

$$X^N(0) \Rightarrow x(0) \tag{5.19}$$

for $N \to \infty$, where

$$x := (\mu e, M + \mu e) \quad \text{and} \quad M := 1 - \mu/\lambda. \tag{5.20}$$

THEOREM 5.1 *Assume that* $\mu < \lambda$. *Then, for* $N \to \infty$,

$$\Gamma(X^N) \Rightarrow \Gamma(x). \tag{5.21}$$

The proof of Theorem 5.1 is given in Sect. 5.5.

The fluid limiting behaviour of the queues in the system is easily characterised from Theorem 5.1 and is given by

$$\Gamma(x)(t) = \begin{pmatrix} 1 - (t/\tau - \lfloor t/\tau \rfloor) \\ t/\tau - \lfloor t/\tau \rfloor \end{pmatrix} M, \tag{5.22}$$

where

$$\tau := M/\mu. \tag{5.23}$$

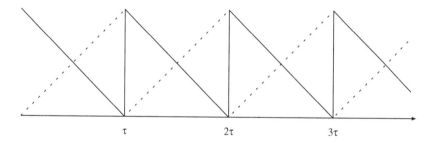

Fig. 5.1. Fluid limiting behaviour of fraction of jobs in ante room $W^N(\cdot)$, *dotted line*, and fraction of jobs at single server queue $Q^N(\cdot)$, *solid line*

Thus the fraction of the jobs in the ante room and the single server queue combined is constant and equal to M. Consequently, the fraction of jobs in the infinite server queue is also constant and equal to $1 - M$. Both the fraction of jobs in the ante room and the fraction of jobs at the single server queue exhibit a periodic behaviour, illustrated in Fig. 5.1, with cycles of fixed length τ. Starting with an empty ante room at the transfer instant, the fraction of the jobs in the ante room increases linearly, until a fraction M of the total number of jobs is in the ante room. At this instant the ante room becomes empty again. The fraction of jobs at the single server queue exhibits the complementary behaviour.

Though the queue sizes have a deterministic fluid limit, this is not the case for the sojourn times, due to the randomness in the service discipline. However, the distribution of the sojourn time can easily be derived from the deterministic behaviour of the queues. Indeed, a job that arrives at the ante room at time t will wait in the ante room until the next transfer instant. Next, it will stay at the single server for a period of time which is uniformly distributed over the, fixed, cycle length, and which is independent of its arrival instant at the ante room.

A result to this end is stated more formally. Let $\hat{F}_b^N$ denote an empirical measure defined by the sojourn times of the jobs arriving in $(\tau_{b-1}, \tau_b]$: $\hat{F}_b^N(x)$ is the number of jobs in batch b with sojourn times less than or equal to x; see (5.47) for a formal definition.

THEOREM 5.2 *For any* $b \geq 1$:

$$\frac{1}{NM} \hat{F}_b^N(x) \Rightarrow \begin{cases} \frac{x^2}{2\tau^2} & \text{for } x \in [0, \tau] \\ 1 - \frac{(2\tau - x)^2}{2\tau^2} & \text{for } x \in [\tau, 2\tau], \end{cases} \tag{5.24}$$

almost surely for $N \to \infty$.

Theorem 5.2 is proven in Sect. 5.5. It states that the empirical distribution of the sojourn times converges to the convolution of two uniform distributions. Thus, Theorem 5.2 corroborates a conjecture about the variance of the sojourn time given in Sect. 4.5 (4.29).

5.4   Transfer Map

In this section, we derive the properties of the transfer map, Γ, defined in (5.18) that are needed for fluid and diffusion approximations.

5.4.1   Measurability of Γ

It is relevant to note that the τ_i can be defined via the operations of composition and left-continuous inverse:

$$\tau_i(d, a) = (d^{\leftarrow} \circ a)(\tau_{i-1}) \qquad (5.25)$$
$$= (d^{\leftarrow} \circ a)^i(0),$$

where

$$(x \circ y)(t) := x(y(t)), \qquad (5.26)$$
$$(x \circ y)^i := (x \circ y)^{i-1} \circ x \circ y, \qquad (5.27)$$

and, see Whitt [172], Remark 13.6.1,

$$(x^{\leftarrow})(t) := \inf(s > 0 : x(s) \geq t). \qquad (5.28)$$

It now follows that $\Gamma(d, a)$ is a well defined and measurable map from $\mathbf{D}_u \times \mathbf{D}_\uparrow$ to $\mathbf{D} \times \mathbf{D}$. This follows from the measurability of addition and composition and the measurability of γ as defined in (5.17). To verify measurability of γ, observe that each τ_i is measurable $\mathbf{D}_u \times \mathbf{D}_\uparrow$ to $\mathbb{R}$ due to the representation (5.25) and the measurability of the left-continuous inverse, composition, and the projection map. This implies measurability of the map $\mathbf{D}_u \times \mathbf{D}_\uparrow$ to $\mathbb{R}^{j^*}$ which maps (d, a) to the sequence $\{\tau_i, i = 0, 1, \dots\}$. The measurability of γ then follows from the representation of γ as a countable sum:

$$\gamma(d, a)(t) := \begin{cases} \sum_{j=0}^{j^*-1} \tau_j \mathbf{1}_{[\tau_j, \tau_{j+1})}(t) & \text{for } t \in [0, \tau^*), \\ \tau^* & \text{for } t \geq \tau^*, \end{cases} \qquad (5.29)$$

and from the measurability of addition and the measurability of the function Θ defined as

$$\Theta(g) := \sum_{j=0}^{J} \mathbf{1}_{[g_j, g_{j+1})}(t) \qquad (5.30)$$

operating on strictly increasing sequences $g = (g_j, j = 0, \dots, J + 1)$, such that the $[g_j, g_{j+1})$ cover $\mathbb{R}^+$.

5.4.2   Continuity of Γ

Define the space $\tilde{\Omega}$ as the subset of function pairs $(d, a) \in \mathbf{C}_{\uparrow\uparrow} \times \mathbf{C}_\uparrow$ for which d satisfies a rate constraint:

$$K_1 s \leq d(t + s) - d(t) \leq K_2 s \qquad (5.31)$$

for fixed positive real-valued constants K_1, K_2, and any $s > 0$, and for which

$$\inf(a - d) > 0. \tag{5.32}$$

LEMMA 5.1 *The transfer map Γ defined in (5.18) is continuous at $(d, a) \in \tilde{\Omega}$.*

Proof Consider $(d, a) \in \tilde{\Omega}$ and a sequence of function pairs $(d^\epsilon, a^\epsilon) \in \mathbf{D}_u \times \mathbf{D}_\uparrow$ indexed by ϵ, such that

$$(d^\epsilon, a^\epsilon) \xrightarrow{J_1} (d, a) \tag{5.33}$$

for $\epsilon \to 0$. We will show that (5.33) implies that

$$\Gamma(d^\epsilon, a^\epsilon) \xrightarrow{J_1} \Gamma(d, a) \tag{5.34}$$

for $\epsilon \to 0$.

Because of the continuity of addition on $\mathbf{C} \times \mathbf{D}$, see Whitt [170], Theorem 4.1, and the continuity of composition of $\mathbf{C} \times \mathbf{D}_\uparrow$, see [170], Theorem 3.1, it follows from the representation (5.18), that it suffices to show that (5.33) implies

$$\gamma(d^\epsilon, a^\epsilon) \xrightarrow{J_1} \gamma(d, a) \tag{5.35}$$

for $\epsilon \to 0$. Hence, it remains to be shown that

$$\lim_{\epsilon \to 0} d_{J_1}^T \left(\gamma(d^\epsilon, a^\epsilon), \gamma(d, a) \right) = 0 \tag{5.36}$$

for all T except a countable set.

For this, use τ_i, $i = 0, 1, \ldots$ as a shorthand for the transfer instants $\tau_i(d, a)$ and τ_i^ϵ, $i = 0, 1, \ldots$ for the transfer instants $\tau_i(d^\epsilon, a^\epsilon)$. Concerning these transfer instants, we prove at the end of this section that

LEMMA 5.2 *For all $T > 0$:*

$$\lim_{\epsilon \to 0} \max_{\tau_i \in [0, T]} |\tau_i - \tau_i^\epsilon| = 0. \tag{5.37}$$

Define h^ϵ as the function that maps τ_i onto τ_i^ϵ, $i = 0, 1, \ldots$, that maps T onto T, and that is linear in between:

$$h^\epsilon(t) := \begin{cases} \frac{\tau_{i+1} - t}{\tau_{i+1} - \tau_i} \tau_i^\epsilon + \frac{t - \tau_i}{\tau_{i+1} - \tau_i} \tau_{i+1}^\epsilon & \text{for } t \in [\tau_i, \tau_{i+1}), \\ \frac{T - t}{T - \tau_i} \tau_i^\epsilon + \frac{t - \tau_i}{T - \tau_i} T & \text{for } t \in [\tau_i, T]. \end{cases} \tag{5.38}$$

Fix $T \notin \text{Disc}(\Gamma(x))$. There is ϵ^* so that h^ϵ is a homeomorphism from $[0, T]$ onto $[0, T]$ for all $\epsilon < \epsilon^*$. Hence, for $\epsilon < \epsilon^*$

$$d_{J_1}^T(\gamma(d, a), \gamma(d^\epsilon, a^\epsilon)) = \inf_{h \in \mathcal{H}_T} ||\gamma(d, a) - \gamma(d^\epsilon, a^\epsilon) \circ h||_T \vee ||h - e||_T$$

$$\leq ||\gamma(d, a) - \gamma(d^\epsilon, a^\epsilon) \circ h^\epsilon||_T \vee ||h^\epsilon - e||_T.$$

For $t \in [\tau_i, \min(\tau_{i+1}, T))$,

$$\gamma(d, a)(t) = \tau_i$$

and

$$\gamma(d^\epsilon, a^\epsilon) \circ h^\epsilon(t) = \tau_i^\epsilon,$$

so that

$$||\gamma(d, a) - \gamma(d^\epsilon, a^\epsilon) \circ h^\epsilon||_T = \max_{\tau_i \in [0, T]} |\tau_i - \tau_i^\epsilon|. \qquad (5.39)$$

Moreover, from the definition (5.38)

$$||h^\epsilon - e||_T = \max_{\tau_i \in [0, T]} |\tau_i - \tau_i^\epsilon|. \qquad (5.40)$$

Consequently,

$$\lim_{\epsilon \to 0} d_{J_1}^T(\gamma(d, a), \gamma(d^\epsilon, a^\epsilon)) = 0$$

for all $T \notin \mathrm{Disc}(\Gamma(x))$ because of Lemma 5.2. $\qquad \square$

Proof of Lemma 5.2. First note that there are only finitely many $\tau_i \in [0, T]$. To verify this claim, define

$$\Delta := \inf_{t \in [0, T]} (a(t) - d(t)) \qquad (5.41)$$

and $\Delta > 0$ by assumption. Hence for all τ_i defined in (5.14) such that $\tau_{i+1} \in [0, T]$, we have

$$
\begin{aligned}
\tau_{i+1} - \tau_i &= \inf_{\delta > 0} (d(\tau_i + \delta) - d(\tau_i) \geq a(\tau_i) - a(\tau_{i-1})) \\
&= \inf_{\delta > 0} (d(\tau_i + \delta) - d(\tau_i) \geq a(\tau_i) - d(\tau_i)) \\
&\geq \inf_{\delta > 0} (d(\tau_i + \delta) - d(\tau_i) \geq \Delta) \\
&\geq \inf_{\delta > 0} (K_2 \delta \geq \Delta) \\
&= \Delta / K_2.
\end{aligned}
$$

It follows that there are at most a finite number of τ_i in each finite interval $[0, T]$.

Next, we prove that τ_i is close to τ_i^ϵ for all i, and, as there are only finitely many $\tau_i \in [0, T]$ we can do so using induction. Now, $\tau_0 = \tau_0^\epsilon = 0$ by definition. Next, assume that

$$\lim_{\epsilon \to 0} |\tau_i - \tau_i^\epsilon| = 0 \qquad (5.42)$$

for some $i \geq 0$. Let $\eta > 0$ be given. We will identify ϵ_0 such that

$$|\tau_{i+1} - \tau^\epsilon_{i+1}| \leq \eta \tag{5.43}$$

for all $\epsilon < \epsilon_0$. For this, consider

$$\tau^\epsilon_{i+1} := \inf_s (d^\epsilon(s) \geq a^\epsilon(\tau^\epsilon_i))$$

$$= \inf_s (d(s) \geq a(\tau_i) + a^\epsilon(\tau^\epsilon_i) - a(\tau^\epsilon_i)$$

$$+ a(\tau^\epsilon_i) - a(\tau_i) - d^\epsilon(s) + d(s)).$$

Because of (5.33), there exists ϵ_1 such that

$$||(d, a) - (d^\epsilon, a^\epsilon)||_T \leq K_1 \eta / 3$$

for all $\epsilon < \epsilon_1$. By uniform continuity of a on $[0, T]$, there exists $\delta > 0$ such that

$$|a(x) - a(y)| \leq K_1 \eta / 3$$

for all $(x, y) \in [0, T] \times [0, T]$ such that $|x - y| \leq \delta$. Moreover, by the induction hypothesis (5.42), there exists ϵ_2 such that

$$|\tau_i - \tau^\epsilon_i| \leq \delta$$

for all $\epsilon < \epsilon_2$. Taking $\epsilon_0 = \epsilon_1 \wedge \epsilon_2$, we find that

$$\tau^\epsilon_{i+1} \leq \inf(d(s) \geq a(\tau_i) + K_1 \eta)$$

$$= \tau_{i+1} + \inf(d(s + \tau_{i+1}) \geq K_1 \eta)$$

$$\leq \tau_{i+1} + \eta,$$

for all $\epsilon < \epsilon_0$, where we have used the rate constraint (5.31). Similarly, we can show that

$$\tau^\epsilon_{i+1} \geq \tau_{i+1} - \eta$$

for ϵ sufficiently small. Now combining this shows (5.43) and this proves the lemma. $\qquad \square$

5.5 Proofs of Theorems

Proof of Theorem 5.1. We first prove that

$$X^N \Rightarrow x \tag{5.44}$$

for $N \to \infty$ using Theorem 2.2 in Mandelbaum et al. [117]. To use this theorem, observe that $A^N(\cdot), D^N(\cdot), Y^N(\cdot)$ can be represented as

$$
\begin{pmatrix} D^N(t) \\ A^N(t) \\ Y^N(t) \end{pmatrix} = \begin{pmatrix} 0 \\ 0 \\ Y^N(0) \end{pmatrix} + \begin{pmatrix} 0 \\ 1 \\ 1 \end{pmatrix} A_1 \left(N \int_0^t \lambda(1 - Y^N(s)/N) \mathrm{d}s \right)
$$
$$
+ \begin{pmatrix} 1 \\ 0 \\ -1 \end{pmatrix} A_2 \left(N \int_0^t \mu_1(Y^N(s) > 0) \mathrm{d}s \right), \tag{5.45}
$$

where A_1 and A_2 are two independent Poisson processes. Now the theorem in [117] states that Assumption (5.19) implies that

$$
\frac{1}{N} \begin{pmatrix} D^N \\ A^N \\ Y^N \end{pmatrix} \Rightarrow \begin{pmatrix} d \\ a \\ y \end{pmatrix} \tag{5.46}
$$

for $N \to \infty$. Here (d, a, y) uniquely solve the integral equation

$$
\begin{pmatrix} d(t) \\ a(t) \\ y(t) \end{pmatrix} = \begin{pmatrix} 0 \\ 0 \\ M \end{pmatrix} + \begin{pmatrix} 0 \\ 1 \\ 1 \end{pmatrix} \left(\int_0^t \lambda(1 - y(s)) \mathrm{d}s \right)
$$
$$
+ \begin{pmatrix} 1 \\ 0 \\ -1 \end{pmatrix} \left(\int_0^t \mu \mathbf{1}(y(s) > 0) \mathrm{d}s \right).
$$

Now (5.44) immediately follows.

Statement (5.21) follows from the generalised continuous-mapping Theorem, see [172], Theorem 3.4.3. The transfer map Γ is well defined and measurable on $\mathbf{D}_u \times \mathbf{D}_\uparrow$ and continuous on $\tilde{\Omega}$, see Theorem 5.1. From (5.44), we verify that the limiting sequence is indeed in $\tilde{\Omega}$ with probability 1. $\square$

Proof of Theorem 5.1. Define

$$
\hat{F}_b^N(x) = \sum_{l=A^N(\tau_{b-1}^N)+1}^{A^N(\tau_b^N)} \mathbf{1}\left(\xi_{D^N(\sigma_l^N)+1}^* + \sum_{i=D^N(\sigma_l^N)+2}^{D^N(\tau_b^N)+U_l^N} \xi_i \le N\mu x \right), \tag{5.47}
$$

where

$$
\xi_{D^N(\sigma_l^N)+1}^* = \sum_{i=1}^{D^N(\sigma_l^N)+1} \xi_i - \sigma_l^N \tag{5.48}
$$

is the residual service time of the job in service during the lth arrival to the ante room, and

$$U_l^N = \sum_{i=A^N(\tau_{b-1}^N)+1}^{A^N(\tau_b^N)} 1(u_i \le u_l) \qquad (5.49)$$

equals the service position in the single server queue for the lth arrival to the ante room in the processing of batch b. Consequently, $\hat{F}_b^N(x)$ is the empirical measure defined by the total sojourn times in ante room and service queue of the jobs in batch b. Next, define

$$\hat{F}_{W,b}^N(x) = \sum_{l=A^N(\tau_{b-1}^N)+1}^{A^N(\tau_b^N)} 1\left(\xi_{D^N(\sigma_l^N)+1}^* + \sum_{i=D^N(\sigma_l^N)+2}^{D^N(\tau_b^N)} \xi_i \le N\mu x \right), \qquad (5.50)$$

and

$$\hat{F}_{Q,b}^N(x) = \sum_{l=D^N(\tau_b^N)+1}^{D^N(\tau_{b+1}^N)} 1\left(\sum_{i=D^N(\tau_b^N)+1}^{D^N(\tau_b^N)+U_l^N} \xi_i \le N\mu x \right), \qquad (5.51)$$

with ξ^* and U_l^N as above. Consequently, $\hat{F}_{W,b}^N$ is the empirical measure of the sojourn times in the ante room of the jobs in batch b, and $\hat{F}_{Q,b}^N$ is the empirical measure of the sojourn times at the single server queue for batch b. We will show that $\hat{F}_b^N$ is the convolution of $\hat{F}_{Q,b}^N$ and $\hat{F}_{W,b}^N$ plus a remainder which vanishes almost surely in the limit. Then we apply Lemma 5.3 below.

Introduce Q_b^N for the number of jobs in batch b:

$$Q_b^N := A^N(\tau_b^N) - A^N(\tau_{b-1}^N),$$

and define

$$a_{k,l} := \sum_{i=D^N(\sigma_l^N)+2}^{D^N(\tau_b^N)+k} \xi_i,$$

$$c_{k,l}(x) := 1\left(\xi_{D^N(\sigma_l^N)+1}^* + a_{k,l} \le N\mu x \right),$$

$$b_{k,l} := 1\left(U_l^N = k \right) - 1/Q_b^N,$$

$$d_k := \sum_{i=D^N(\tau_b^N)+1}^{D^N(\tau_b^N)+k} \xi_i,$$

and rewrite $\hat{F}_b^N$:

$$\hat{F}_b^N(x) = \sum_{l=A^N(\tau_{b-1}^N)+1}^{A^N(\tau_{b-1}^N)+Q_b^N} \sum_{k=1}^{Q_b^N} \mathbf{1}\left(\xi_{D^N(\sigma_l^N)+1}^* + a_{k,l} \le N\mu x\right) \mathbf{1}\left(U_l^N = k\right)$$

$$= \sum_{l=A^N(\tau_{b-1}^N)+1}^{A^N(\tau_{b-1}^N)+Q_b^N} \sum_{k=1}^{Q_b^N} c_{k,l}(x)\mathbf{1}\left(U_l^N = k\right)$$

$$= \sum_{l=A^N(\tau_{b-1}^N)+1}^{A^N(\tau_{b-1}^N)+Q_b^N} \sum_{k=1}^{Q_b^N} c_{k,l}(x)\frac{1}{Q_b^N} + \zeta^N(x),$$

where

$$\zeta^N(x) = \sum_{l=A^N(\tau_{b-1}^N)+1}^{A^N(\tau_{b-1}^N)+Q_b^N} \sum_{k=1}^{Q_b^N} c_{k,l}(x)b_{k,l}.$$

Observe that

$$c_{k,l}(x) = \mathbf{1}\left(\xi_{D^N(\sigma_l^N)+1}^* + a_{0,l} + d_k \le N\mu x\right),$$

so that

$$\hat{F}_b^N(x) = \sum_{k=1}^{Q_b^N} \hat{F}_{W,b}^N\left(x - d_k/N\mu\right)\frac{1}{Q_b^N}d\hat{F}_{Q,b}^N\left(d_k/N\mu\right) + \zeta^N(x).$$

Thus, we have expressed $\hat{F}_b^N/Q_b^N$ as the convolution of two distribution functions plus a remainder:

$$\frac{1}{Q_b^N}\hat{F}_b^N = \frac{1}{Q_b^N}\hat{F}_{W,b}^N \star \frac{1}{Q_b^N}\hat{F}_{Q,b}^N + \frac{1}{Q_b^N}\zeta^N(x). \tag{5.52}$$

Now as we shall prove that

$$\lim_{N\to\infty} \frac{1}{N}\zeta^N(x) = 0 \tag{5.53}$$

almost surely, the statement (5.24) of Theorem 5.2 immediately follows from the observation that

$$\lim_{N\to\infty} F_1^N \star F_2^N = F_1 \star F_2, \tag{5.54}$$

if for $i = 1, 2$:

$$\lim_{N\to\infty} F_i^N = F_i \tag{5.55}$$

and Lemma 5.3, below.

Finally, to prove (5.53), note that

$$\text{var}\left(\frac{1}{N}\zeta^N(x)\right) = \mathbb{E}\left(\left(\frac{1}{N}\zeta^N(x)\right)^2\right).$$

On squaring $\zeta^N(x)$, observing that $|c_{k,l}(x)| \leq 1$,

$$Q_b^N \to \infty$$

for $N \to \infty$, and

$$\mathbb{E}\left(b_{k,l}b_{k',l'}\right) = \begin{cases} 0 & k \neq k' \ l \neq l' \\ -\left(\frac{1}{Q_b^N}\right)^2 & k \neq k' \ l = l' \\ -\left(\frac{1}{Q_b^N}\right)^2 & k = k' \ l \neq l' \\ \frac{1}{Q_b^N}\left(1 - \frac{1}{Q_b^N}\right) & k = k' \ l = l', \end{cases}$$

it follows that

$$\lim_{N\to\infty} \text{var}\left(\frac{1}{N}\zeta^N(x)\right) = 0,$$

so that (5.53) holds. □

LEMMA 5.3 *For $b \geq 1$ and any $x \in [0, \tau]$ we have that*

$$\frac{1}{N}\hat{F}_{W,b}^N(x) \Rightarrow \frac{x}{\tau}M \quad \text{as} \quad N \to \infty, \tag{5.56}$$

and

$$\frac{1}{N}\hat{F}_{Q,b}^N(x) \Rightarrow \frac{x}{\tau}M \quad \text{as} \quad N \to \infty. \tag{5.57}$$

Proof To prove (5.56), define

$$\hat{G}_{b,1}^N(x) := \tau_b^N - \inf\left(t \in [\tau_{b-1}^N, \tau_b^N) : \frac{1}{N\mu}\sum_{i=D^N(t)+1}^{D^N(\tau_b^N)} \xi_i \leq x\right) \tag{5.58}$$

and

$$\hat{G}_{b,2}^N(x) := \tau_b^N - \inf\left(t \in [\tau_{b-1}^N, \tau_b^N) : \frac{1}{N\mu}(D^N(\tau_b^N) - D^N(t)) \leq x\right). \tag{5.59}$$

We will show that for any $\delta > 0$ and $x \geq 0$:

$$\lim_{N\to\infty} \mathbb{P}(|\frac{1}{N}\hat{F}_{W,b}^N(x) - \mu\hat{G}_{b,1}^N(x)| > \delta) = 0 \tag{5.60}$$

and that

$$\lim_{N\to\infty} \mathbb{P}(|\hat{G}_{b,1}^N(x) - \hat{G}_{b,2}^N(x)| > \delta) = 0, \tag{5.61}$$

so that (5.56) follows by combining the fluid limit for D^N with (5.59).

To prove (5.60), note that σ_l^N is increasing in l, and that

$$\xi_{D^N(\sigma_l^N)+1}^* \le \xi_{D^N(\sigma_l^N)+1}.$$

Consequently,

$$\xi_{D^N(\sigma_l^N)+1}^* + \sum_{i=D^N(\sigma_l^N)+1}^{D^N(\tau_b^N)} \xi_i$$

is decreasing in l. Hence, there is a first, smallest, l for which the condition

$$\xi_{D^N(\sigma_l^N)+1}^* + \sum_{i=D^N(\sigma_l^N)+2}^{D^N(\tau_b^N)} \xi_i \le N\mu x$$

is satisfied. Denote this first value by $l^*(x)$:

$$l^*(x) := \inf\left(A^N(\tau_{b-1}^N) + 1 \le l \le A^N(\tau_b^N) : \xi_{D^N(\sigma_l^N)+1}^* + a_{0,l} \le N\mu x\right),$$

and observe that

$$\hat{F}_{W,b}^N(x) = A^N(\tau_b^N) - l^*(x) + 1. \tag{5.62}$$

Next, define $t^*(x)$ as the value of t for which the infimum in (5.58) is attained

$$t^*(x) := \inf\left(\tau_{b-1}^N \le t < \tau_b^N : \frac{1}{N\mu}\sum_{i=D^N(t)+1}^{D^N(\tau_b)} \xi_i \le x\right). \tag{5.63}$$

Then, by definition,

$$\sigma_{l^*(x)-1}^N < t^*(x) \le \sigma_{l^*(x)}^N. \tag{5.64}$$

Hence,

$$\frac{1}{N}\hat{F}_{W,b}^N(x) - \mu\hat{G}_{b,1}^N(x) \tag{5.65}$$

$$= (A^N(\tau_b^N)/N - \mu\tau_b^N) - (l^*(x)/N - \mu t^*(x)) + 1/N$$

$$\le (A^N(\tau_b^N)/N - \mu\tau_b^N) - (l^*(x)/N - \mu\sigma_{l^*(x)}^N) + 1/N$$

$$= (A^N(\tau_b^N)/N - \mu\tau_b^N) - (A^N(\sigma_{l^*(x)}^N)/N - \mu\sigma_{l^*(x)}^N) + 1/N,$$

where the last equality follows from the identity $A^N(\sigma_l^N) = l$ for all l. From the fluid limit for A^N we get $\frac{1}{N}A^N \Rightarrow \mu e$ for $N \to \infty$ uniformly on compact sets, it follows that the upper bound in (5.65) vanishes almost surely. A similar argument, using the lower bound in (5.64), shows that a lower bound for (5.65) vanishes almost surely and this completes the proof of (5.60).

To show (5.61), note that

$$\frac{1}{N\mu} \sum_{i=D^N(\sigma_l^N)+1}^{D^N(\tau_b^N)} \xi_i = \frac{D^N(\tau_b^N) - D^N(\sigma_l^N)}{N} + \eta_l^N, \qquad (5.66)$$

where

$$\eta_l^N = \frac{1}{N\mu} \sum_{i=D^N(\sigma_l^N)+1}^{D^N(\tau_b^N)} (\xi_i - E(\xi_i)), \qquad (5.67)$$

with $\eta_l^N \Rightarrow 0$ for all l for $N \to \infty$.

Next, to show (5.57), note that

$$\hat{F}_{Q,b}^N(x) \overset{d}{=} \sum_{l=D^N(\tau_b^N)+1}^{D^N(\tau_{b+1}^N)} 1 \left(\frac{1}{N\mu} \sum_{i=D^N(\tau_b^N)+1}^{l} \xi_i \leq x \right), \qquad (5.68)$$

and proceed as in the proof of the fluid limit for $\hat{F}_{W,b}^N$. $\qquad \square$

5.6 Conclusion

In this chapter, we have derived fluid limits for the queue-size processes in a closed queueing system with the GROS discipline. We have done this by a mapping approach. Additionally, we have analysed the limit of the empirical measure associated with the sojourn times of the jobs in batch b, in a way that uses the fluid limits of the queue-size processes.

Several topics remain for further research. Firstly, we have concentrated on the mapping approach and the necessary properties for the map that make this approach work, giving less attention to the limits of the basic processes that are mapped. In fact, we obtain scaling limits for the basic departure and arrival processes, under restrictive exponential conditions, from a general theorem in Mandelbaum et al. [117]. The references [75] and [103] show that the assumptions can be relaxed considerably, and the results in this chapter carry over without additional effort to the more general model where one makes assumptions only about the first two moments of the service time distributions.

Secondly, it is relevant to obtain further refinements to the fluid approximations of the queue-size processes via diffusion approximations, see, e.g. [172], Internet Supplement Chap. 9, or Mandelbaum and Massey [116] and Mandelbaum and Pats [118]. In fact, it seems very well possible to apply the theory in [172] to also obtain the diffusion limits of the queue-size processes as the arrival and departure processes are continuous. However, complications arise as the image under the mapping is not continuous as a function of time. These discontinuities are a consequence of the gate openings which characterise the service discipline of the system displayed in Fig. 4.2. These gate openings cause an instantaneous transfer of jobs from the ante room to the single server queue and this gives rise to discontinuities in the fluid approximations to the number of jobs in the ante room and the number of jobs in the queue. Thus we are faced with the situation of providing a diffusion approximation in case the fluid limit is discontinuous and in which the convergence to the fluid limit uses the J_1-metric.

This topic is taken up in Denteneer and Gromoll [49] who compare the standard approach to diffusion approximations with an alternative. The standard approach considers scaled differences between the limiting sequence and its fluid limit. In [49] it is shown by example that this approach may break down if the convergence to the fluid limit uses the J_1-metric and an alternative approach is proposed, based on scaled differences between the time-perturbed limiting sequence and the fluid limit. It is shown that this procedure is non-unique in that the diffusion limit itself may depend on the details of the time perturbations. However, it is also proven that the total approximation to the original process, combining the fluid and diffusion limit, is unique. An interesting further topic, not considered in [49], is to determine the limit of the empirical measure associated with the sojourn times of the jobs in batch b, in a way that uses the diffusion limits of the queue-size processes.

Thirdly, in Sect. 4.2.1 we have described the use of a scheduled blocked access scheme that improves on the standard blocked access protocol. Moreover, in Sect. 4.6 we have introduced the repairman model with the Gated Partial Random Order of Service (GPROS) discipline as an appropriate model for trees with scheduled blocked access. The generalisation of the results in this chapter to the case of GPROS is an interesting topic for further study. Here, it seems that the techniques presented in this chapter are very well applicable to the case in which a gate period is split into a fixed number of intervals. However, the case in which this number of intervals increases with the number of machines is particularly challenging.

PART III

BULK SERVICE

Chapter 6

METHODOLOGY

In Sect. 2.3 we proposed variants of the bulk service queue as models for the data-transmission phase of a reservation procedure. In this chapter we first focus on the classical bulk service queue and, in particular, on deriving characteristics of the stationary queue length distribution. The bulk service queue has a deeply rooted place in queueing theory and appeared throughout the twentieth century in a variety of applications. The work done on the bulk service queue runs to a large extent parallel to the maturing of queueing theory as a branch of mathematics. We therefore give an extensive description of the historical perspective in which the bulk service queue can be placed. Next, we give a detailed account of the methodology that can be applied to solve for the stationary queue length distribution. The methodology can be roughly categorised into three techniques: The generating function technique, random walk theory, and the Wiener–Hopf technique. Depending on the technique used, characteristics of the stationary distribution can be expressed in terms of either the roots of some equation, or infinite series that involve convolutions of some probability distribution. The three techniques cover the existing methodology to a large extent, both from the analytical and computational viewpoint. The historical overview is given in Sect. 6.1. We then present the generating function technique in Sect. 6.2, random walk theory in Sect. 6.3, and the Wiener–Hopf technique in Sect. 6.4.

6.1 Historical Perspective

The first, somewhat disguised, appearance of the discrete bulk service queue was in the theory of telephone exchanges, going by the name $M/D/s$ queue. This model was introduced in the 1920s by Erlang, see [25], who is considered to be the founding father of queueing theory. At a telephone exchange with s available channels, calls arrive according to a Poisson process. Each call occupies a channel for a constant holding time. Let X_n denote the number of calls, both waiting and in service just after the nth holding time. Then, the following relation holds:

$$X_{n+1} = (X_n - s)^+ + A_n, \qquad (6.1)$$

where $x^+ = \max(0, x)$ and A_n denotes the number of newly arriving calls during the nth holding time. It should be noted that due to the assumption of constant holding times, the calls which are in progress at the end of the nth holding time must have started during this holding time. Also, the calls which terminate during the nth holding time must have started before the beginning of this holding time.

The random variables A_n, $n = 0, 1, \ldots$ are assumed to be i.i.d. according to a random variable A that has a Poisson distribution. Under the assumption that $\mathbb{E}A < s$, the stationary distribution of the Markov chain defined by (6.1) exists.

Erlang obtained expressions for both the first moment and the distribution function of the stationary waiting time for values of $s = 1, 2, 3$. A first formal proof was derived by Crommelin [41] in 1932, although this had already been indicated by Erlang. Crommelin used the generating function technique, which was remarkable at such an early stage, to obtain the pgf of X expressed in terms of the s roots on and within the unit circle of $z^s = \exp(\lambda(z - 1))$. From this pgf, Crommelin could obtain the distribution function of the stationary waiting time. At about the same time, Pollaczek treated the $M/D/s$ queue in a series of papers, generalising it to the $M/G/s$ queue. Pollaczek's work [137] was difficult to read, since he relied on rather complicated analysis, so Crommelin [42] gave an exposition of Pollaczek's theory for the $M/D/s$ queue and found his own results in agreement with those of Pollaczek. Both methods lead to a solution in terms of infinite series that involve convolutions of the Poisson distribution. It is noteworthy that, after a lull in the literature of more than sixty years, Franx [68] came up recently with alternative expressions for the stationary waiting time distribution in the $M/D/s$ queue.

The infinite series-type result was generalised by Pollaczek [138]. In his derivation of the stationary waiting time distribution for the $G/G/1$ queue, Pollaczek obtained an identity, which was some years later obtained independently and by a different method by Spitzer [156]. Pollaczek again used complicated analysis, whereas Spitzer gave an elegant combinatorial proof. This is probably the reason why the result goes down in history as Spitzer's identity, despite the efforts of Syski [158], who pointed out the equivalence of the two results. For a detailed treatment of Spitzer's identity, we refer to Sect. 6.3.

6.1.1 From Telephony to Digital Data Transfer

Recursion (6.1) that describes the queue length process in the $M/D/s$ queue fits into the framework of bulk service queues. In this type of queues, at each epoch of service, a number of customers is taken from the queue. The bulk service queue originates from the work of Bailey [9] in 1954. Bailey modelled the situation where a doctor is prepared to see a maximum of no more than s patients per clinic session. The new patients who arrive during the clinic session join the queue right after the session ends. Bailey assumed that patients arrive according to a Poisson process, and in case of deterministic visiting times (Bailey allows for generally distributed visiting times) the recursive relation (6.1) would hold. Note that both the $M/D/s$ queue and Bailey's bulk service queue are continuous-time models that can be described in terms of discrete random variables by assuming Poisson arrivals and considering the queue at specific (embedded) points in time.

The first real discrete-time bulk service queue was introduced by Boudreau et al. [20] in 1962. They modelled the situation of a helicopter leaving a station every twenty minutes carrying a maximum of s passengers. Passengers that arrive between subsequent departures join the queue just after the next departure instant, again leading to (6.1), except now A can be any discrete random variable (with $\mathbb{E}A$ smaller than s), instead of just Poisson. This generalisation does not increase the complexity much, and so the method applied by Boudreau et al. [20] is almost identical to that of Bailey.

Up till the mid 1970s, applications of bulk service queues were scarce. The most notable exception is the problem of estimating delays at traffic lights that alternate between periods of red and green (yellow is disregarded) of fixed length. For this traffic problem, bulk service queueing theory has been used to develop closed-form approximations for the expected delay, see, e.g. Darroch [43], McNeill [121], Miller [125], Newell [134] and Webster [169].

The real resurrection of the interest in the bulk service queue came in the mid 1970s with the emergence of computer applications and digital data transfer. During the last decades of the twentieth century, discrete-time models have been applied to model digital communication systems such as multiplexers and packet switches. In this field, the discrete bulk service queue plays a key role due to its wide range of applications, among which the Asynchronous Transfer Mode (ATM) switching element, see Bruneel and Kim [28] and the references therein. For this model, time is divided into slots of fixed length, and again (6.1) holds with X_n the queue content, in terms of packets, at the beginning of slot n, A_n the number of new packets that arrive during slot n, and s the maximum number of packets that can be served during one slot. Besides the discrete bulk service queue, there are many other types of bulk queueing models, for which we refer to Baghi and Templeton [8], Bruneel and Kim [28], Cohen [35], Chaudhry and Templeton [32] and Powell [142].

6.1.2   Methodology

Deriving expressions for Laplace–Stieltjes transforms or pgf's that contain roots of some equation has become a classic procedure in queueing theory. When applying the generating function technique, as introduced by Crommelin [41], the consideration of roots is often inevitable. Initially, the need for roots was considered to be a slur on the transform solutions, since the determination of the roots could be numerically troublesome and the roots themselves have no probabilistic interpretation. However, due to advanced numerical algorithms and increased computational power, root-finding has become more or less straightforward. In Chaudhry et al. [31] it is demonstrated that root-finding in queueing theory is well structured, in the sense that the roots are distinct for most models and that their location is well predictable, so that numerical problems are not likely to occur.

In case of the discrete bulk service queue, there is at least one alternative to root-finding. Using the recursive relation (6.1), the distribution of X_{n+1} follows from the convolution of the distribution of $(X_n - s)^+$ and the distribution of A. Since discrete convolutions are not so hard to compute, see, e.g. Ackroyd [3], one could iterate (6.1) to obtain the transient queue length distributions which eventually will tend to the stationary distribution for increasing values of n. This idea of iterating (6.1) can be made more rigorous using random walk, or fluctuation, theory.

Many of the results from random walk theory are important for queueing theory. In particular, the waiting-time process in the $G/G/1$ queue where customers are served in order of arrival can be viewed as a random walk with a reflecting barrier at zero. The evolution equation that relates the waiting times of two subsequent customers is nowadays referred to as Lindley's equation and given by

$$W_{n+1} = (W_n + B_n - C_n)^+, \quad n = 0, 1, \ldots, \tag{6.2}$$

where W_n denotes the waiting time of the nth arriving customer, B_n denotes the service time of the nth arriving customer, and C_n denotes the interarrival time between the nth and $(n+1)$st arriving customer. Lindley [114] showed that, due to the $\max(0, \cdot)$ operator, finding the stationary waiting-time distribution requires the solution of a Wiener–Hopf type integral equation. With these observations, Lindley opened up a new field of research in which the Wiener–Hopf technique, see, e.g. Smith [155] or De Smit [154], and other methods from random walk theory were used to study queueing models. For many types of queues, the Wiener–Hopf technique leads to an explicit factorisation in terms of the roots of some characteristic equation. For an overview of the results from random walk theory that play a role in queueing theory we refer to Cohen [35], Sect. I.6.6, and Asmussen [7], Chap. 8. Perhaps the most famous result is the earlier-mentioned Spitzer's identity which, among other things, expresses the Laplace transform of the stationary waiting-time distribution in terms of an infinite series that involves convolutions of some given probability distribution, see Sect. 6.3 for a detailed treatment.

It is quite common that for a particular queueing model, one or more of the processes of interest may be described in terms of a Lindley equation. In fact, (6.1) is a Lindley equation as well. This means that the methods developed to solve Lindley's equation for the general case become also available for the discrete bulk service queue. Equation (6.1) allows for a Wiener–Hopf factorisation, which results in the same solution for the pgf of the stationary queue length as obtained with the generating function technique. Again, the solution requires the roots of some characteristic equation.

We have mentioned three techniques that can be applied to deal with the discrete bulk service queue: The generating function technique, random walk theory and the Wiener–Hopf technique. All three techniques can be applied to solve for the stationary regime and result in the pgf of the stationary queue length, denoted by $X(z)$. The generating function technique is the most traditional method and leads to an expression for $X(z)$ that includes the roots on and inside the unit circle of some equation. Random walk theory comes into the picture when one observes that the queue length process is a random walk with a reflecting barrier at zero. Spitzer's identity then yields an expression for $X(z)$ in terms of infinite series that involve convolutions of the probability distribution of A. The Wiener–Hopf technique allows for two solutions: $X(z)$ in terms of roots as obtained by the generating function technique and $X(z)$ in terms of infinite series as obtained from random walk theory. In that respect we might say that the Wiener–Hopf technique can be considered as the broadest approach. However, its application is far from straightforward and requires more advanced mathematics than is needed for the generating function technique and random walk theory. Therefore, we first present the latter two techniques, and then derive the same results with the Wiener–Hopf technique. Although the three techniques each have a broad range of applications, we present them, for reasons of clarity, in the context of the discrete bulk service queue.

6.2 Generating Function Technique

The discrete bulk service queue is defined by the recursion

$$X_{n+1} = (X_n - s)^+ + A_n. \tag{6.3}$$

Here, time is assumed to be slotted, X_n denotes the queue length at the beginning of slot n, A_n denotes the number of new packets that arrive during slot n, and s denotes the maximum number of packets that can be transmitted in one slot. Packets that arrive to the queue in slot n can be transmitted at the earliest from the beginning of slot $n + 1$. This is no restrictive assumption, since studying the queue $X_{n+1} = (X_n + A_n - s)^+$ is equivalent, see Sect. 6.3.

We denote for a non-negative discrete random variable Y its mean by $\mathbb{E}Y$ or μ_Y, its variance by σ_Y^2 and $\mathbb{P}(Y = j)$ by y_j. Furthermore, we denote the pgf of Y by $Y(z)$, i.e., $Y(z) = \sum_{j=0}^{\infty} y_j z^j$, which is known to be analytic for $|z| < 1$ and continuous for $|z| \leq 1$. The number of new packets that arrives per slot is assumed to be i.i.d. according to a discrete random variable A with $a_j = \mathbb{P}(A = j)$ and pgf $A(z)$. We assume that $a_0 > 0$, which involves no essential limitation: If a_0 were zero, we would replace the distribution $\{a_i\}_{i\geq 0}$ by $\{a_i'\}_{i\geq 0}$ where $a_i' = a_{i+m}$, a_m being the first non-zero entry of $\{a_i\}_{i\geq 0}$, and a corresponding decrease in the maximum number of packets transmitted per slot according to $s' = s - m$.

Assume that $\mu_A < s$. Then, the stationary queue length distribution exists, see, e.g. Bruneel and Kim [28]. Let X denote the random variable following the stationary distribution of the Markov chain defined by (6.3), with

$$x_j = \mathbb{P}(X = j) = \lim_{n \to \infty} \mathbb{P}(X_n = j), \quad j = 0, 1, 2, \ldots. \tag{6.4}$$

The stationary queue length distribution satisfies the balance equations

$$x_k = \sum_{j=s}^{s+k} x_j a_{k-j+s} + \sum_{j=0}^{s-1} x_j a_k, \quad k = 0, 1, 2, \ldots. \tag{6.5}$$

Multiplying both sides of the above expression with z^k and summing over all values of k yields

$$
\begin{aligned}
X(z) &= \sum_{k=0}^{\infty} x_k z^k \\
&= \sum_{k=0}^{\infty} \sum_{j=s}^{s+k} x_j a_{k-j+s} z^k + \sum_{k=0}^{\infty} \sum_{j=0}^{s-1} x_j a_k z^k \\
&= z^{-s} \sum_{j=s}^{\infty} x_j z^j \sum_{k=j-s}^{\infty} a_{k-j+s} z^{k-j+s} + \sum_{j=0}^{s-1} x_j \sum_{k=0}^{\infty} a_k z^k \\
&= z^{-s} X(z) A(z) - z^{-s} \sum_{j=0}^{s-1} x_j z^j A(z) + \sum_{j=0}^{s-1} x_j A(z). \tag{6.6}
\end{aligned}
$$

Rewriting (6.6) results in the following expression for $X(z)$, see, e.g. Bruneel and Kim [28],

$$X(z) = \frac{A(z) \sum_{j=0}^{s-1} x_j (z^s - z^j)}{z^s - A(z)}, \quad |z| \leq 1. \tag{6.7}$$

The expression (6.7) is of indeterminate form, but the s unknowns $x_0, \ldots,$ x_{s-1} can be determined by consideration of the zeros of the denominator in (6.7) that lie on or within the unit circle, see, e.g. Bailey [9] or Zhao and Campbell [176].

We can prove the following result:

THEOREM 6.1 *If $\mu_A < s$ and $a_0 > 0$, the equation $z^s = A(z)$ has s roots on or within the unit circle.*

Proof See Appendix 6.A. □

The s roots of $z^s = A(z)$ in $|z| \leq 1$ are denoted by $z_0 = 1, z_1, \ldots, z_{s-1}$. For the ease of presentation we assume that these roots are distinct, but the theory presented below can be easily extended to the case in which there are multiple roots, see Remark 6.1.

Since the function $X(z)$ is finite on and inside the unit circle, the numerator of the right-hand side of (6.7) needs to be zero for each of the s roots, i.e., the numerator should vanish at the exact points where the denominator of the right-hand side of (6.7) vanishes. This gives the following s equations

$$\sum_{j=0}^{s-1} x_j(z_k^s - z_k^j) = 0, \quad k = 0, 1, \ldots, s - 1. \tag{6.8}$$

For $z_0 = 1$, the above equation has a trivial solution, but the normalisation condition $X(1) = 1$ provides an additional equation. Using l'Hôpital's rule, this equation is found to be

$$s - \mu_A = \sum_{j=0}^{s-1} x_j(s - j), \tag{6.9}$$

where both sides represent the average unused service capacity.

The system of equations can be written in matrix form $\mathbf{Ax} = \mathbf{b}$, where $\mathbf{x}$ denotes the column vector $(x_0, x_1, \ldots, x_{s-1})^T$, and $\mathbf{b}$ the column vector with all entries zero except for the first entry which is equal to $s - \mu_A$. The matrix $\mathbf{A}$ is given by

$$\mathbf{A} = \begin{pmatrix} s & s-1 & \cdots & 1 \\ z_1^s - 1 & z_1^s - z_1 & \cdots & z_1^s - z_1^{s-1} \\ z_2^s - 1 & z_2^s - z_2 & \cdots & z_2^s - z_2^{s-1} \\ \vdots & \vdots & \vdots & \vdots \\ z_{s-1}^s - 1 & z_{s-1}^s - z_{s-1} & \cdots & z_{s-1}^s - z_{s-1}^{s-1} \end{pmatrix}. \tag{6.10}$$

For this system of s equations to have a unique solution, all s equations should be linearly independent. Denote the determinant of a matrix $\mathbf{C}$ as $|\mathbf{C}|$. For the case that the roots $z_0 = 1, z_1, \ldots, z_{s-1}$ are distinct Bailey [9] has shown that $|\mathbf{A}| = |\mathbf{V}|$, where $\mathbf{V}$ is some Vandermonde matrix. In that case, $\mathbf{A}$ is non-singular and a unique solution $x_0, x_1, \ldots, x_{s-1}$ exists. Using some additional arguments, we can derive explicit expressions for the x_j as given in the following lemma:

LEMMA 6.1 *If the roots* $z_0 = 1, z_1, \ldots, z_{s-1}$ *are distinct, the set of equations* (6.8) *together with the normalisation condition* (6.9) *constitute a system of s linearly independent equations. The unique solution is given by*

$$x_j = (-1)^{j+2}(s - \mu_Y)\frac{S_{s-j} + S_{s-j-1}}{\prod_{k=1}^{s-1}(z_k - 1)}, \quad j = 0, 1, \ldots, s - 1, \quad (6.11)$$

where S_j denotes the elementary symmetric function of degree j, having as variables $z_1, \ldots, z_{s-1}$, *i.e.*

$$S_j = \sum_{1 \le i_1 < i_2 < \cdots < i_j \le s-1} z_{i_1} z_{i_2} \ldots z_{i_j}. \quad (6.12)$$

Proof See Appendix 6.C. $\qquad\qquad\qquad\qquad\qquad\qquad\qquad\qquad\qquad\qquad\qquad$ □

REMARK 6.1 If one (or more) of the roots $z^s = A(z)$ in $|z| \le 1$ has multiplicity higher than 1, an expression like (6.11) for the x_j cannot be derived. However, for the pgf $X(z)$ to be finite on and inside the unit circle, the numerator of (6.7) should still have the same zeros as the denominator of (6.7), and with the same multiplicity. For $z_0 = 1$ it can be verified that this root has multiplicity 1, and we have argued before that this root places no restriction on the probabilities $x_0, \ldots, x_{s-1}$ whatsoever. For all other roots, the fact that the numerator of (6.7) should vanish does yield a restriction on $x_0, \ldots, x_{s-1}$. Assume, for example, that z_1 has multiplicity 2. Then z_1 should be a double root of the numerator of (6.7), yielding next to (6.8),

$$\sum_{j=0}^{s-1} x_j(sz_1^{s-1} - jz_1^{j-1}) = 0, \quad (6.13)$$

as an additional restriction on $x_0, \ldots, x_{s-1}$. In a similar way, whatever the multiplicity of the roots would be, we can construct $s - 1$ equations. Together with the normalisation equation (6.9) this gives s equations for s unknowns. Since the Markov chain has a unique stationary distribution, we know that this system of equations has a unique solution.

So we can determine the probabilities $x_0, \ldots, x_{s-1}$ either explicitly through (6.11), or implicitly through a system of linear equations as described in

Remark 6.1. From these probabilities, the entire probability distribution can be found. That is, from matching coefficients at both sides of

$$(z^s - A(z))X(z) = A(z) \sum_{j=0}^{s-1} x_j(z^s - z^j), \tag{6.14}$$

we find that

$$x_j = \frac{1}{a_0}\left(x_{j-s} - a_{j-s}\sum_{n=0}^{s-1} x_n - \sum_{n=0}^{j-s-1} x_{s+n}a_{j-s-n}\right), \quad j \geq s. \tag{6.15}$$

6.2.1 Roots On and Inside the Unit Circle

We can go a step further and eliminate $x_0, \ldots, x_{s-1}$ from (6.7). Write

$$\sum_{j=0}^{s-1} x_j(z^s - z^j) = \gamma_1(z-1)\prod_{k=1}^{s-1}(z - z_k), \tag{6.16}$$

where the constant γ_1 can be determined from differentiating both sides of (6.16) with respect to z, and using the normalisation condition (6.9). This gives

$$\gamma_1 = \frac{s - \mu_A}{\prod_{k=1}^{s-1}(1 - z_k)}, \tag{6.17}$$

and so

$$\sum_{j=0}^{s-1} x_j(z^s - z^j) = (s - \mu_A)(z-1)\prod_{k=1}^{s-1}\frac{z - z_k}{1 - z_k}. \tag{6.18}$$

Together with (6.7) this yields the following result:

THEOREM 6.2 *The pgf of the stationary queue length distribution is given by*

$$X(z) = \frac{A(z)(z-1)(s - \mu_A)}{z^s - A(z)}\prod_{k=1}^{s-1}\frac{z - z_k}{1 - z_k}, \quad |z| \leq 1. \tag{6.19}$$

Explicit expressions for the mean μ_X and variance σ_X^2 of the stationary queue length can be obtained by evaluating derivatives of $X(z)$ at $z = 1$, i.e. $\mu_X = X'(1)$ and $\sigma_X^2 = X''(1)+X'(1)-X'(1)^2$. This gives, see, e.g. Laevens and Bruneel [105],

$$\mu_X = \frac{\sigma_A^2}{2(s - \mu_A)} + \frac{1}{2}\mu_A - \frac{1}{2}(s-1) + \sum_{k=1}^{s-1}\frac{1}{1 - z_k}, \tag{6.20}$$

$$\sigma_X^2 = \sigma_A^2 + \frac{A'''(1) - s(s-1)(s-2)}{3(s - \mu_A)} + \frac{A''(1) - s(s-1)}{2(s - \mu_A)}$$
$$+ \left(\frac{A''(1) - s(s-1)}{2(s - \mu_A)}\right)^2 - \sum_{k=1}^{s-1}\frac{z_k}{(1 - z_k)^2}. \tag{6.21}$$

6.2.2   Roots Outside the Unit Circle

When A has finite support, i.e. $A \leq m$, we know that $A(z)$ is a polynomial of degree m. It then immediately follows that $z^s = A(z)$ has $m - s$ roots outside the unit circle, to be denoted by $z_s, z_{s+1}, \ldots, z_{m-1}$, and so we can write (with $m > s$)

$$\frac{\sum_{j=0}^{s-1} x_j(z^s - z^j)}{z^s - A(z)} = \frac{\gamma_2 \prod_{k=0}^{s-1}(z - z_k)}{\prod_{k=0}^{m-1}(z - z_k)}$$

$$= \frac{\gamma_2}{\prod_{k=s}^{m-1}(z - z_k)}, \tag{6.22}$$

where γ_2 is a constant. From the normalisation condition $X(1) = 1$ it follows that $\gamma_2 = \prod_{k=s}^{m-1}(1 - z_k)$, and so we arrive at

THEOREM 6.3 *The pgf of the stationary queue length distribution is given by*

$$X(z) = A(z) \prod_{k=s}^{m-1} \frac{1 - z_k}{z - z_k}, \quad |z| \leq 1. \tag{6.23}$$

From (6.23) we obtain, see, e.g. Zhao and Campbell [176],

$$\mu_X = \mu_A + \sum_{k=s}^{m-1} \frac{1}{z_k - 1}, \tag{6.24}$$

$$\sigma_X^2 = \sigma_A^2 + \sum_{k=s}^{m-1} \frac{1}{(z_k - 1)^2} + \sum_{k=s}^{m-1} \frac{1}{z_k - 1}. \tag{6.25}$$

Using partial-fraction expansion, see, e.g. Henrici [79], inverting $X(z)$ is a simple exercise. Write

$$\prod_{k=s}^{m-1} \frac{1 - z_k}{z - z_k} = \sum_{i=s}^{m-1} \frac{r_i}{z - z_i}, \tag{6.26}$$

where

$$r_i = \lim_{z \to z_i} (z - z_i) \prod_{k=s}^{m-1} \frac{1 - z_k}{z - z_k}$$

$$= \frac{\prod_{k=s}^{m-1}(1 - z_k)}{\prod_{k=s, k \neq i}^{m-1}(z_i - z_k)}, \quad i = s, \ldots, m - 1. \tag{6.27}$$

Then rewrite the right-hand side of (6.26) as

$$\sum_{i=s}^{m-1} \frac{r_i}{z - z_i} = -\sum_{n=0}^{\infty} \sum_{i=s}^{m-1} \left(\frac{r_i}{z_i}\right) \left(\frac{1}{z_i}\right)^n z^n. \tag{6.28}$$

Hence, the probability distribution $\{x_j\}_{j=0}^{\infty}$ is given by

$$x_j = -\sum_{n=0}^{j}\sum_{i=s}^{m-1}\left(\frac{r_i}{z_i}\right)\left(\frac{1}{z_i}\right)^{j-n}a_n, \quad j = 0, 1, 2, \ldots. \quad (6.29)$$

REMARK 6.2 For j large enough, the sum on the right-hand side of (6.29) is dominated by the pole of $X(z)$ with the smallest modulus, to be denoted by $\hat{z}$. This pole can be shown to be the unique root of $z^s = A(z)$ contained in the interval $(1, \infty)$, see, e.g. Tijms [161]. Omitting all fractions in (6.29) other than the one that corresponds to $\hat{z}$ gives the following approximation for the tail probabilities:

$$x_j \approx -\sum_{n=0}^{j}\left(\frac{\hat{r}}{\hat{z}}\right)\left(\frac{1}{\hat{z}}\right)^{j-n}a_n, \quad j \to \infty. \quad (6.30)$$

6.3 Random Walk Theory

Most results from random walk theory that are important for queueing theory have been presented in the context of the waiting time of a customer in the $G/G/1$ queue, see, e.g. Asmussen [7], Chap. 10. We first show that the discrete bulk service queue may be viewed as a special type of $G/G/1$ queue. Then we invoke a result for the $G/G/1$ queue known as *Spitzer's identity* that leads to an alternative expression for the pgf of the stationary queue length in the discrete bulk service queue.

The discrete bulk service queue is closely related to the discrete $D/G/1$ queue. The latter refers to a single server queue at which customers arrive with discrete and deterministic interarrival times, are served on a first-come-first-served basis and have service requirements that are i.i.d. according to a discrete random variable A. The waiting time of the nth customer, denoted by W_n, then satisfies, see, e.g. Servi [152],

$$W_{n+1} = (W_n + A_n - s)^+, \quad n = 0, 1, \ldots. \quad (6.31)$$

Here, A_n denotes the service time of customer n and the integer s denotes the interarrival time between two consecutive customers. When $\mathbb{E}A < s$, the stationary waiting time denoted by W exists, see, e.g. Servi [152]. By comparing (6.31) and (6.3), it is immediately clear that the pgf's of the stationary distributions of the discrete bulk service queue and the discrete $D/G/1$ queue are related as $X(z) = A(z)W(z)$. Hence, a solution for $W(z)$ yields the solution for $X(z)$ and vice versa.

From the evolution equation (6.31) it can be seen that the distribution of W_{n+1} follows from the convolution of the distribution of W_n and that of

$A_n - s$, corrected for the $\max(0, \cdot)$ operator. Hence, by iterating on (6.31) one can obtain transient characteristics of the model. This idea of iterating can be made more rigorous using random walk theory. When we set W_0 equal to zero, the following result is known as Spitzer's identity, see Spitzer [156], p. 207:

THEOREM 6.4 (Spitzer's identity) *For $0 \leq t < 1$, $|z| \leq 1$,*

$$\sum_{n=0}^{\infty} t^n \mathbb{E}(z^{W_n}) = \exp\left(\sum_{l=1}^{\infty} t^l l^{-1} \mathbb{E}(z^{S_l^+})\right), \qquad (6.32)$$

where $S_l = \sum_{i=1}^{l} (A_i - s)$, A_i i.i.d. as A.

From (6.32) the distribution of the stationary waiting time W can be obtained. When we write (6.32) as

$$(1-t)\sum_{n=0}^{\infty} t^n \mathbb{E}(z^{W_n}) = \exp\left(\sum_{l=1}^{\infty} t^l l^{-1} \mathbb{E}(z^{S_l^+} - 1)\right), \qquad (6.33)$$

it follows from Abel's theorem, see [156], p. 207, or Cohen [35], p. 650, that $W(z)$ is given by

$$W(z) = \lim_{t \uparrow 1}(1-t)\sum_{n=0}^{\infty} t^n \mathbb{E}(z^{W_n}) = \exp\left(\sum_{l=1}^{\infty} l^{-1} \mathbb{E}(z^{S_l^+} - 1)\right). \quad (6.34)$$

Now we return to the discrete bulk service queue. The pgf of the stationary queue length is given by $X(z) = A(z)W(z)$ with $W(z)$ as in (6.34), which gives the following result.

THEOREM 6.5 *The pgf of the stationary queue length distribution is given by*

$$X(z) = A(z)\exp\left(\sum_{l=1}^{\infty} \frac{1}{l}\mathbb{E}(z^{S_l^+} - 1)\right), \quad |z| \leq 1, \qquad (6.35)$$

where $S_l = \sum_{i=1}^{l} (A_i - s)$, A_i i.i.d. according to A.

The mean and variance of the stationary queue length follow from taking derivatives of (6.34). Note that

$$W'(1) = \left[\sum_{l=1}^{\infty} \frac{1}{l}\mathbb{E}(S_l^+ z^{S_l^+ - 1})W(z)\right]_{z=1} = \sum_{l=1}^{\infty} \frac{1}{l}\mathbb{E}(S_l^+), \qquad (6.36)$$

$$W''(1) = \sum_{l=1}^{\infty}\sum_{k=1}^{\infty} \frac{1}{l}\mathbb{E}(S_l^+)\frac{1}{k}\mathbb{E}(S_k^+) + \sum_{l=1}^{\infty} \frac{1}{l}\mathbb{E}(S_l^+(S_l^+ - 1)). \qquad (6.37)$$

Denoting by A^{*l} the random variable that follows the l-fold convolution of the distribution of A, this gives after some rewriting

$$\mu_X = \mu_A + \sum_{l=1}^{\infty} \frac{1}{l} \sum_{j=ls}^{\infty} (j - ls)\mathbb{P}(A^{*l} = j), \qquad (6.38)$$

$$\sigma_X^2 = \sigma_A^2 + \sum_{l=1}^{\infty} \frac{1}{l} \sum_{j=ls}^{\infty} (j - ls)^2 \mathbb{P}(A^{*l} = j), \qquad (6.39)$$

which are root-free expressions for μ_X and σ_X^2, and alternative expressions for (6.20)–(6.21) and (6.24)–(6.25).

Moreover, introducing the short-hand notation $C_{z^j}[f(z)]$ for the coefficient of z^j in $f(z)$, the following result follows from (6.35):

LEMMA 6.2 *The stationary queue length distribution is given by (for $j = 0, 1, \ldots$)*

$$x_j = \mathbb{P}(W = 0) \sum_{k=0}^{j} a_k \, C_{z^{j-k}} \left[\exp \left(\sum_{l=1}^{\infty} \sum_{i=1}^{\infty} \frac{1}{l} \mathbb{P}(A^{*l} = ls + i) z^i \right) \right],$$

$$(6.40)$$

and

$$\mathbb{P}(W = 0) = \exp \left(- \sum_{l=1}^{\infty} \sum_{i=ls+1}^{\infty} \frac{1}{l} \, \mathbb{P}(A^{*l} = i) \right). \qquad (6.41)$$

Expression (6.40) provides for each x_j a root-free representation, as an alternative for (6.11), (6.15) and (6.29) that do depend on the roots of $z^s = A(z)$. For determining the coefficients C_{z^j} in (6.40), the following property can be used:

PROPERTY 6.1 For $K(z) = \sum_{j=0}^{\infty} k_j z^j$ and $M(z) = \sum_{j=0}^{\infty} m_j z^j$ with $K(z) = \exp(M(z))$, the coefficients k_j follow recursively from the coefficients m_j, and vice versa, according to

$$k_0 = \exp(m_0); \quad k_j = \frac{1}{j} \sum_{n=1}^{j} n m_n k_{j-n}, \quad j = 1, 2, \ldots. \qquad (6.42)$$

Proof The proof consists of computing the k_j's successively by equating coefficients in $K'(z) = M'(z)K(z)$. □

REMARK 6.3 Several authors, e.g. Konheim [102], Murata and Miyahara [128], and Stadje [157], have suggested to approximate the $G/G/1$ queue by its discrete counterpart. This can be done as follows. Denote by B_n the service time of customer n and by C_n the interarrival time between customer n and

$n + 1$. Choose B_n and C_n i.i.d. according to discrete random variables B and C, respectively. Moreover, assume $C \leq s$. Then W_n satisfies

$$W_{n+1} = (W_n + B_n - C_n)^+ = (W_n + A_n - s)^+, \quad n = 0, 1, \ldots, \quad (6.43)$$

with A_n assumed i.i.d. as $A = B - C + s$. The discrete approximation to the $G/G/1$ queue then fits into the framework of the $D/G/1$ queue.

6.4   Wiener–Hopf Technique

The Wiener–Hopf technique stems from mathematical physics, and found its way to the field of applied probability through Smith [155], Kemperman [94] and Cohen [34, 35], also see Regterschot [147]. Perhaps the most famous application of the Wiener–Hopf technique is in the context of random walks, see, e.g. Cohen [35] or Asmussen [7]. As the Wiener–Hopf technique is a powerful tool for the analysis of Markov processes whose evolution equation contains the $\max(0, \cdot)$ operator, it can also be applied in the case of the discrete bulk service queue. We will apply the Wiener–Hopf technique to obtain alternative derivations of the expressions for $X(z)$ given by (6.19), (6.23) and (6.35). In addition to this application, the Wiener–Hopf technique is also frequently applied in the analysis of the trajectories of random walks, see, e.g. Asmussen [7].

Let us first describe the role of the $\max(0, \cdot)$ operator. From recursion (6.3) we have

$$
\begin{aligned}
\mathbb{E}(z^{X_{n+1}}) &= \mathbb{E}(z^{A_n} \mathbf{1}(X_n \leq s)) + \mathbb{E}(z^{X_n + A_n - s} \mathbf{1}(X_n > s)) \\
&= \mathbb{P}(X_n \leq s)\mathbb{E}(z^{A_n}) + \mathbb{E}(z^{X_n + A_n - s}) \\
&\quad - \mathbb{E}(z^{X_n + A_n - s} \mathbf{1}(X_n \leq s)), \quad (6.44)
\end{aligned}
$$

where $\mathbf{1}(x) = 1$ if x is true and 0 otherwise. Letting $n \to \infty$ and observing that X_n and A_n are independent then yields

$$\xi_+(z)(1 - z^{-s}A(z)) = \xi_-(z), \quad (6.45)$$

where $\xi_+(z) = X(z)/A(z)$ and $\xi_-(z) = \mathbb{P}(X \leq s) - \mathbb{E}(z^{X-s}\mathbf{1}(X \leq s))$. Observe that ξ_+ (respectively ξ_-) is analytic and bounded in $|z| < 1$ (respectively $|z| > 1$), and both ξ_+, ξ_- are continuous up to $|z| = 1$.

In order to find an explicit expression for $\xi_+(z)$ we need to factorise the function $1 - z^{-s}A(z)$. In more general terms, we need to factorise a function $1 - Y(z)$, where $Y(z)$ is the pgf of a random variable Y for which it holds that $\mathbb{E}Y < 0$ (in the case of the discrete bulk service queue Y is the difference of A and s, i.e. $Y(z) = z^{-s}A(z)$). Such a factorisation is known as the *Wiener–Hopf factorisation*. The treatment of this factorisation in terms of a characteristic function prevails in the literature, but we present the theory here for $Y(z)$ being a pgf; Bayer [11] does this also. Furthermore, it is common practice to present a factorisation of the bivariate function $1 - rY(z)$, $0 \le r < 1$, but since we are interested in the stationary distribution only (and not in the transient distribution), we will stick to the analysis of the univariate function $1 - Y(z)$. The Wiener–Hopf factorisation identity then reads, see Asmussen [7], p. 228, or Prabhu [146], p. 22:

THEOREM 6.6 (Wiener–Hopf factorisation identity) *The following decomposition exists:*

$$1 - Y(z) = \phi_+(z)\phi_-(z), \quad |z| = 1, \quad (6.46)$$

where ϕ_+ (respectively ϕ_-) is analytic and bounded in $|z| < 1$ (respectively $|z| > 1$), and both ϕ_+, ϕ_- are continuous up to $|z| = 1$.

Hence, once we know the functions ϕ_+, ϕ_- we can write (6.45) as

$$\xi_+(z)\phi_+(z) = \frac{\xi_-(z)}{\phi_-(z)}, \quad (6.47)$$

where the left-hand side (respectively right-hand side) of (6.47) represents a function that is analytic and bounded in $|z| < 1$ (respectively $|z| > 1$), and both sides of (6.47) are functions continuous up to $|z| = 1$. Therefore, their analytic continuation contains no singularities in the entire complex plane. Liouville's theorem then says

THEOREM 6.7 (Liouville) *Let $f(z)$ be analytic for all values of z and let $|f(z)| < K$ for all values of z, where K is a constant (so that $|f(z)|$ is bounded as $|z| \to \infty$). Then $f(z)$ is seen to be constant.*

Whence upon using Liouville's theorem the left-hand side of (6.47) is constant, and since $\xi_+(1) = 1$, we obtain

$$\xi_+(z) = \frac{\phi_+(1)}{\phi_+(z)}. \quad (6.48)$$

With the machinery described above, we present alternative proofs of Theorems 6.2, 6.3 and 6.5, where we rely on three different factorisations of the function $1 - Y(z)$.

Alternative proof of Theorem 6.5. Start from the basic identity

$$1 - z = \exp(\ln(1 - z)) = \exp\left(-\sum_{l=1}^{\infty} \frac{1}{l} z^l\right), \quad |z| \le 1, \; z \ne 1. \quad (6.49)$$

We have denoted $\sum_{i=1}^{l}(A_i - s)$ by S_l for which it holds that $\mathbb{E}(z^{S_l}) = (z^{-s}A(z))^l$ and $|z^{-s}A(z)| < 1$ for $|z| = 1$. Hence, we can write (for $|z| = 1$)

$$1 - z^{-s}A(z) = \exp\left(-\sum_{l=1}^{\infty} \frac{1}{l}(z^{-s}A(z))^l\right) = \phi_+(z)\phi_-(z), \quad (6.50)$$

where

$$\phi_+(z) = \exp\left(-\sum_{l=1}^{\infty}\frac{1}{l}\mathbb{E}(z^{S_l}\mathbf{1}(S_l > 0))\right), \tag{6.51}$$

$$\phi_-(z) = \exp\left(-\sum_{l=1}^{\infty}\frac{1}{l}\mathbb{E}(z^{S_l}\mathbf{1}(S_l \leq 0))\right). \tag{6.52}$$

Observe that

$$\phi_+(1) = \exp\left(-\sum_{l=1}^{\infty}\frac{1}{l}\mathbb{P}(S_l > 0)\right), \tag{6.53}$$

which by (6.48) completes the proof. □

The type of Wiener–Hopf factorisation as outlined above can be applied for the more general $G/G/1$ queue in a similar fashion. For a subclass of queues, the Wiener–Hopf technique allows for an explicit factorisation that relies on the consideration of the roots of some equation, see, e.g. Asmussen [7], Cohen [35], or Kleinrock [98]. For the discrete bulk service queue, the knowledge on the roots of $z^s = A(z)$ given by Theorem 6.1 can be applied to prove the following two previously derived results:

Alternative proof of Theorem 6.2. We construct an explicit factorisation of $1 - z^{-s}A(z)$ by choosing

$$\phi_+(z) = \frac{z^s - A(z)}{\prod_{k=0}^{s-1}(z - z_k)}, \quad \phi_-(z) = \frac{\prod_{k=0}^{s-1}(z - z_k)}{z^s}. \tag{6.54}$$

With

$$\phi_+(1) = \lim_{z \to 1}\frac{z^s - A(z)}{(z - 1)\prod_{k=1}^{s-1}(z - z_k)} = \frac{s - \mu_A}{\prod_{k=1}^{s-1}(1 - z_k)}, \tag{6.55}$$

this completes the proof. □

Alternative proof of Theorem 6.3. In case $A \leq m$, we construct an explicit factorisation of $1 - z^{-s}A(z)$ by choosing

$$\phi_+(z) = \gamma\prod_{k=s}^{m-1}(z - z_k), \quad \phi_-(z) = \frac{\prod_{k=0}^{s-1}(z - z_k)}{z^s}, \tag{6.56}$$

with γ a constant. We have that

$$\phi_+(1) = \gamma\prod_{k=s}^{m-1}(1 - z_k), \tag{6.57}$$

which completes the proof. □

6.5   Summary

The generating function technique is widely applied in the discrete-time modelling of communication systems. The required procedures of numerical root-finding are known to a broad community of engineers, including electrical engineers and computer scientists. On the contrary, the Wiener–Hopf technique and random walk theory, while often of great interest to theoreticians, are less popular among practitioners, probably due to the advanced mathematical techniques involved and the lack of clearly described computational schemes for determining certain performance characteristics. For the discrete bulk service queue, this separation between theoretical and practical results is far less clear-cut. The results obtained by all three methods in fact might complement each other well, since clear descriptions of the computational schemes are available.

Depending on the method used, one can obtain a transform solution of the stationary queue length distribution either in terms of the roots of $z^s = A(z)$ or in terms of infinite series that involve convolutions of the distribution of A. The solution in terms of roots can be obtained using either the generating function technique or the Wiener–Hopf technique. The solution in terms of the infinite series follows from random walk theory (Spitzer's identity), and can also be obtained using the Wiener–Hopf technique. All four options have been demonstrated in this chapter. To make a rough distinction, two courses can be followed in obtaining characteristics of the stationary queue length distribution of the discrete bulk service queue: roots or infinite series. That is, for the mean, variance and probability distribution we have the expressions in terms of roots and the expressions in terms of infinite series. From a practical viewpoint, both courses have their difficulties: Roots need to be determined and infinite series need to be truncated.

In Janssen and van Leeuwaarden [87], the expressions in terms of infinite series are derived from the expressions in terms of roots using Fourier sampling. In Janssen and van Leeuwaarden [88], analytic representations of the roots are presented for a large class of distributions. The infinite series are further investigated in Janssen and van Leeuwaarden [89]. In particular, the infinite series should be truncated, and measures are constructed to characterise their speed of convergence. In Denteneer et al. [50] sharp bounds are derived for the mean and variance of the stationary queue length. The bounds are in closed form and hold for a general arrival process.

APPENDIX 6.A: Proof of Theorem 6.1

This appendix is based on Adan et al. [4]. Rouché's theorem is the standard tool for proving Theorem 6.1, since it is typically used to determine regions of the complex plane in which there may be zeros of a given analytic function. We focus on the zeros of the function $z^s - A(z)$ (i.e. the roots of $z^s = A(z)$) on or within the unit circle.

Rouché's theorem is a direct consequence of the *argument principle* and the scope of application of Rouché's theorem goes well beyond the field of queueing theory. While the verification of the conditions needed to apply Rouché's theorem can become rather difficult, in queueing theory this is usually straight-forward. For most queueing applications, the region of interest is typically the unit disk, and the ingredient that makes Rouché's theorem work is oftentimes the stability condition. This is why Rouché's theorem is a popular and stan-dardised tool in queueing theory.

However, in order to apply Rouché's theorem it is required that $A(z)$ has a radius of convergence larger than 1, see Lemma 6-6.A.3, which is not true in general. A pgf obeys all the rules of power series with non-negative coef-ficients, and since $A(1) = 1$ the radius of convergence of a pgf is at least 1. The shoe thus pinches for those pgf's for which the radius of convergence is exactly 1, some examples are given at the end of this section.

In papers like Bruneel [27], Darroch [43], Powell and Humblet [143], Servi [152] and Zhao and Campbell [176], the assumption is made that $A(z)$ has a radius of convergence larger than 1, so that Rouché's theorem can be applied. In this chapter, we impose no such restriction on $A(z)$. Instead of excluding those functions $A(z)$ with radius of convergence 1, we present a proof of The-orem 6.1 that does not rely on Rouché's theorem and holds for general $A(z)$.

Several other authors proved similar generalisations. In [2], Abolnikov and Dukhovny apply the so-called *generalised principle of the argument*, that was proven by Gakhov et al. [71] in 1973, to give a proof for general $A(z)$. Klimenok [100] extended this result to a larger class of functions, so including $z^s - A(z)$, again using the generalised principle of the argument. An alternative approach to deal with general $A(z)$ was presented by Boudreau et al. [20]. Un-der the condition that all zeros in the unit disk are distinct, they were able to ap-ply the *implicit function theorem* to prove the existence of the zeros. However, examples can be constructed for which there are multiple zeros, and so this approach does not cover the issue in full generality. The key idea of Boudreau et al. is to study the parameterised function $z^s - tA(z)$, $0 \le t < 1$, and then letting t tend to one. The same idea, without making the assumption of distinct zeros, has been used by Gail et al. [70] for a larger class of functions, including $z^s - A(z)$. We present an elementary proof of the existence of the zeros for general $A(z)$ using the classical argument principle and truncation of $A(z)$.

We first describe the classical application of Rouché's theorem, and subse-quently give our proof for general $A(z)$.

Classical Setting

Let us first state Rouché's theorem, see, e.g. Titchmarsh [162]:

THEOREM 6-6.A.8 (Rouché) *Let the bounded region D have as its bound-ary a simple closed contour C. Let $f(z)$ and $g(z)$ be analytic both in D and* on C. Assume that $|f(z)| < |g(z)|$ on C. Then $f(z) - g(z)$ has in D the same number of zeros as $g(z)$, all zeros counted according to their multiplicity.

When $A(z)$ has a radius of convergence larger than one, we can prove the following result concerning the number of zeros on and within the unit circle of $z^s - A(z)$ by using Rouché's theorem:

LEMMA 6-6.A.3 *Let $A(z)$ be a pgf that is analytic in $|z| \leq 1 + \nu$, $\nu > 0$. Assume that $A'(1) < s$, $s \in \mathbb{N}$. Then the function $z^s - A(z)$ has exactly s zeros in $|z| \leq 1$.*

Proof Define the functions $f(z) := A(z)$, $g(z) := z^s$. Because $f(1) = g(1)$ and $f'(1) = A'(1) < s = g'(1)$, we have, for sufficiently small $\epsilon > 0$,

$$f(1 + \epsilon) < g(1 + \epsilon). \tag{6.A.1}$$

Consider all z with $|z| = 1 + \epsilon$. By the triangle inequality and (6.A.1) we have that

$$|f(z)| \leq \sum_{j=0}^{\infty} a_j |z|^j = f(1 + \epsilon) < g(1 + \epsilon) = |g(z)|, \tag{6.A.2}$$

and hence $|f(z)| < |g(z)|$. Because both $f(z)$ and $g(z)$ are analytic for $|z| \leq 1 + \epsilon$, Rouché's theorem tells us that $g(z)$ and $f(z) - g(z)$ have the same number of zeros in $|z| \leq 1 + \epsilon$. Letting ϵ tend to zero yields the proof. □

New Setting
Before we present our main result, we first prove a result on the number and location of zeros of $z^s - A(z)$ on the unit circle. We define the period p of a series $\sum_{-\infty}^{\infty} b_j z^j$ as the largest integer for which $b_j = 0$ whenever j is not divisible by p.

LEMMA 6-6.A.4 *Let $A(z)$ be a pgf of some nonnegative discrete random variable with $A(0) > 0$. Assume $A(z)$ is differentiable at $z = 1$ and $A'(1) < s$, where s is a positive integer. If $z^s - A(z)$ has period p, then $z^s - A(z)$ has exactly p zeros on the unit circle given by the pth roots of unity $\tau_k = \exp(2\pi i k/p)$, $k = 0, 1, \ldots, p - 1$. In each of these zeros, the derivative of $z^s - A(z)$ does not vanish.*

Proof Obviously, any zero ξ of $z^s - A(z)$ with $|\xi| = 1$ is simple, since $|A'(\xi)| \leq A'(|\xi|) = A'(1) < s$ and, thus, $s\xi^{s-1} - A'(\xi) \neq 0$. Furthermore, for any z with $|z| = 1$, $|A(z)| = A(1)$ iff $z^k = 1$ whenever $a_k > 0$. This easily follows from the fact that $|a_0 + a_k z^k| < a_0 + a_k$ if $z^k \neq 1$. So, for z with $|z| = 1$ and $A(z) - z^s = 0$ it follows that $z^k = 1$ for all k with $a_k > 0$, and $z^s = 1$. This implies that $z^p = 1$, which completes the proof. □

Note that the requirement $a_0 = A(0) > 0$ involves no essential limitation: If a_0 were zero we would replace the distribution $\{a_i\}_{i \geq 0}$ by $\{a'_i\}_{i \geq 0}$ where $a'_i = a_{i+m}$, a_m being the first non-zero entry of $\{a_i\}_{i \geq 0}$, and a corresponding decrease in s according to $s' = s - m$.

We are now in a position to give the main result:

THEOREM 6-6.A.9 *Let $A(z)$ be a pgf of some nonnegative discrete random variable with $A(0) > 0$. Assume $A(z)$ is differentiable at $z = 1$ and $A'(1) < s$, where s is a positive integer. Also, let $z^s - A(z)$ have period p. Then the function $z^s - A(z)$ has p zeros on the unit circle given by $\tau_k = \exp(2\pi i k/p)$, $k = 0, 1, \ldots, p - 1$ and exactly $s - p$ zeros in $|z| < 1$.*

Proof Lemma 6-6.A.4 tells us that $F(z) = z^s - A(z)$ has p equidistant zeros *on* the unit circle, and so it remains to prove that this function has exactly $s - p$ zeros *within* the unit circle. Thereto, define, for $N \in \mathbb{N}$, the truncated pgf

$$A_N(z) = \sum_{j=0}^{N-1} a_j z^j, \qquad (6.A.3)$$

where N is a multiple of p. Then $F_N(z) = z^s - A_N(z)$ has obviously s zeros in $z \in D = \{z \in \mathbb{C} : |z| \leq 1\}$, since $A_N(z)$ is a polynomial satisfying $A'_N(1) < s$, and Lemma 6-6.A.3 thus applies. By Lemma 6-6.A.4 we know that $F_N(z)$ has p simple and equidistant zeros on the unit circle. We further have that

$$|A(z) - A_N(z)| \leq 2 \sum_{j=N}^{\infty} a_j, \quad |z| \leq 1, \qquad (6.A.4)$$

$$|A'(z) - A'_N(z)| \leq 2 \sum_{j=N}^{\infty} j a_j, \quad |z| \leq 1. \qquad (6.A.5)$$

Thus, $A_N(z)$ and $A'_N(z)$ *converge uniformly* to $A(z)$ and $A'(z)$ on $z \in D$, respectively. Moreover, if $G : D \to \mathbb{C}$ is continuous, then $G(A_N(z))$ is uniformly convergent to $G(A(z))$ on $z \in D$.

Let z on $C = \{z \in \mathbb{C} : |z| = 1\}$. If for all $n \in \mathbb{N}$ there is a $z_n \in D$ with $0 < |z - z_n| < \frac{1}{n}$ and $F(z_n) = 0$, then $F(z) = 0$ and

$$F'(z) = \lim_{n \to \infty} \frac{F(z_n) - F(z)}{z_n - z} = 0. \qquad (6.A.6)$$

However, this is impossible by Lemma 6-6.A.4. Hence, there is an $\eta > 0$ such that $F(\xi) \neq 0$ for all $\xi \in D(z, \eta) := \{\xi \in D : 0 < |\xi - z| < \eta\}$. Since C is compact, it can be covered by finitely many $D(z, \eta)$'s. Hence, there is a $0 < r < 1$ such that $F(z)$ has no zeros in $r \leq |z| < 1$.

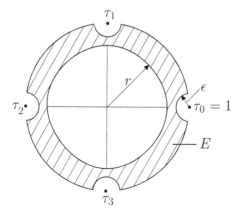

Fig. 6.A.1. Graphical representation of the compact set E

Now we prove that for large N the function $F_N(z)$, as the function $F(z)$, has no zeros in $r \leq |z| < 1$. Thereto, we show that there is an $\epsilon > 0$ and $M \in \mathbb{N}$ such that $F_N(z) \neq 0$ for all $N \geq M$ and $0 < |z - \tau_k| < \epsilon$, $k = 0, 1, \ldots, p-1$. Because $F'(z)$ is continuous and $F'_N(z)$ converges uniformly to $F'(z)$ on $z \in D$, there are $\epsilon > 0$ and $M \in \mathbb{N}$ such that (for $k = 0, 1, \ldots, p-1$)

$$|F'_N(z) - F'(\tau_k)| < \delta < |F'(\tau_k)|, \quad 0 < |z - \tau_k| < \epsilon, \quad N \geq M. \quad (6.A.7)$$

Furthermore, we have (for $k = 0, 1, \ldots, p-1$)

$$|F_N(z) - F'(\tau_k)(z - \tau_k)| = \left| \int_{[\tau_k, z]} (F'_N(s) - F'(\tau_k)) \mathrm{d}s \right|, \quad (6.A.8)$$

where the integration is carried out along the straight line that connects τ_k and z. Hence, for $0 < |z - \tau_k| < \epsilon$ and $N \geq M$, we obtain (for $k = 0, 1, \ldots, p-1$)

$$\left| \int_{[\tau_k, z]} (F'_N(s) - F'(\tau_k)) \mathrm{d}s \right| \leq |z - \tau_k| \max_{s \in [\tau_k, z]} |F'_N(s) - F'(\tau_k)| < |z - \tau_k| \delta.$$

$$(6.A.9)$$

So, it follows that for $0 < |z - \tau_k| < \epsilon$ and $N \geq M$ (for $k = 0, 1, \ldots, p-1$)

$$\begin{aligned}
|F_N(z)| &= |F_N(z) - F'(\tau_k)(z - \tau_k) + F'(\tau_k)(z - \tau_k)| \\
&\geq |F'(\tau_k)||z - \tau_k| - |F_N(z) - F'(\tau_k)(z - \tau_k)| \\
&> (|F'(\tau_k)| - \delta)|z - \tau_k| > 0.
\end{aligned}$$

Since $F_N(z)$ converges uniformly to $F(z)$ and $F(z) \neq 0$ on the compact set, see Fig. 6.A.1,

$$E = \{z \in \mathbb{C} : r \leq |z| \leq 1\} \setminus \bigcup_{k=0}^{p-1} D(\tau_k, \epsilon), \quad (6.A.10)$$

there exists an $K \in \mathbb{N}$ such that $F_N(z) \neq 0$ for all $N \geq K$ and $z \in \mathbb{C}$ with $r \leq |z| < 1$. Hence, for all $N \geq K$ the number of zeros of $F_N(z)$ with $|z| < r$ is equal to $s - p$. This number can be expressed by the argument principle, see, e.g. Titchmarsh [162], as follows

$$s - p = \frac{1}{2\pi i} \oint_{|z|=r} \frac{F_N'(z)}{F_N(z)} dz. \tag{6.A.11}$$

The integrand converges uniformly to $F'(z)/F(z)$, and thus

$$\frac{1}{2\pi i} \oint_{|z|=r} \frac{F'(z)}{F(z)} dz = \lim_{N \to \infty} \frac{1}{2\pi i} \oint_{|z|=r} \frac{F_N'(z)}{F_N(z)} dz = s - p. \tag{6.A.12}$$

Hence, the number of zeros of $F(z)$ with $|z| < r$ is also $s - p$. This completes the proof. □

Obviously, Theorem 6-6.A.9 proves Theorem 6.1. Due to Theorem 6-6.A.9, the $A(z)$ with a radius of convergence of 1 do not have to be excluded from the analysis of the zeros of $z^s - A(z)$. This further means that these pgf's can be incorporated in the general formulation of the solution to the queueing models of interest. The $A(z)$ that have radius of convergence 1 are typically those associated with heavy-tailed random variables. Some examples are given below.

(1) The discrete Pareto distribution, e.g. Johnson et al. [90], defined by

$$a_j = c \frac{1}{j^{p+1}}, \quad j = 1, 2, \dots, \tag{6.A.13}$$

with

$$c = \left(\sum_{j=1}^{\infty} a_j \right)^{-1} = \zeta(p+1)^{-1}, \tag{6.A.14}$$

where $\zeta(\cdot)$ is called the *Riemann zeta function* and $p > 1$. For $k < p$, the kth moment μ_k of the discrete Pareto distribution is given by

$$\mu_k = \frac{\zeta(p - k + 1)}{\zeta(p + 1)}, \tag{6.A.15}$$

whereas for $k \geq p$ the moments are infinite. The discrete Pareto distribution is also known as the Zipf or Riemann zeta distribution.

(2) The discrete standard lognormal distribution, defined by

$$a_j = c e^{-\frac{(\log j)^2}{2}}, \quad j = 1, 2, \dots, \tag{6.A.16}$$

where c is a normalisation constant.

(3) The discrete distribution, related to the continuous Weibull distribution, defined by

$$a_j = cp^{-\sqrt{j}}, \quad j = 0, 1, \ldots, \tag{6.A.17}$$

where $p > 1$ and c is a normalisation constant.

(4) The Haight's zeta distribution, see, e.g. Johnson et al. [90], defined by (with $p > 1$)

$$a_j = \frac{1}{(2j-1)^p} - \frac{1}{(2j+1)^p}, \quad j = 1, 2, \ldots. \tag{6.A.18}$$

APPENDIX 6.B: Numerical Issues

Back to the Roots

The applicability of the generating function approach indisputably depends on finding the roots of $z^s = A(z)$ on and inside the unit circle, because these are needed to determine the unknowns $x_0, x_1, \ldots, x_{s-1}$ in (6.7). Because this issue of root-finding goes a long way back in queueing theory, it has often been addressed, both from analytical and numerical perspectives. We now give a short overview of this root-finding for the Poisson case $A(z) = \exp(\lambda(z-1))$, $\lambda < s$, and point out where extensions can be made to other distributions of A.

The easiest way to determine the roots in the Poisson case is to apply successive substitution to a fixed-point equation. We know that the s roots of $z^s = A(z)$ in $|z| \leq 1$ satisfy

$$z = wA(z)^{1/s} = w\exp(\lambda(z-1)/s), \tag{6.B.1}$$

$w^s = 1$. For each feasible w, (6.B.1) can be shown to have one unique root in $|z| \leq 1$. Moreover, the equations can be solved by successive substitutions as

$$z_k^{(n+1)} = w_k A(z_k^{(n)})^{1/s}, \quad k = 0, 1, \ldots, s-1, \tag{6.B.2}$$

where $w_k = \exp(2\pi i k/s)$, $i = \sqrt{-1}$, and starting values $z_k^{(0)} = 0$. It can be shown that the fixed-point equations (6.B.2) converge to the desired roots. Adan and Zhao [5] distinguish a class of compound Poisson distributions for which the method works. For more general discrete distributions, the method is further investigated in Janssen and van Leeuwaarden [88].

For the Poisson case, an exact description of the roots can be obtained as well. In [88] it is shown, using the Lagrange inversion theorem, that the roots are given by

$$z_k = \sum_{l=1}^{\infty} e^{-l\theta} \frac{(l\theta)^{l-1}}{l!} w_k^l, \quad k = 0, 1, \ldots, s-1, \tag{6.B.3}$$

where $\theta = \lambda/s$. One could truncate the infinite series over l in (6.B.3) to determine the roots. For a large class of discrete distributions, exact expressions for the roots, similar to (6.B.3), are derived in [88].

Although the class of distributions of A for which one can derive an exact expression such as (6.B.3) is far larger than the class for which the method of successive substitutions (6.B.2) works, see [88], neither method works for all distributions. Therefore, the most general method relies on numerical techniques. Chaudhry et al. [31] have developed an application to solve root-finding problems in queueing theory numerically. In our experience, this application works for almost all distributions.

Inversion of a pgf

For the inversion of a pgf we use a technique of Abate and Whitt [1] that relies on the Fourier series method. A distribution $p_0, p_1, \ldots,$ can be retrieved from its pgf $P(z) = \sum_{k=0}^{\infty} p_k z^k$ via

$$p_k = \frac{1}{2\pi i} \oint_{C_r} \frac{P(z)}{z^{k+1}} dz, \qquad (6.B.4)$$

where $i = \sqrt{-1}$ and C_r is a circle about the origin of radius r, $0 < r < 1$. Abate and Whitt [1] approximate (6.B.4) by

$$\hat{p}_k = \frac{1}{2kr^k} \sum_{j=1}^{2k} (-1)^j \mathrm{Re}(P(re^{ij\pi/k})), \qquad (6.B.5)$$

and derive for $0 < r < 1$, $k \geq 1$ the following error bound

$$|p_k - \hat{p}_k| \leq \frac{r^{2k}}{1 - r^{2k}}. \qquad (6.B.6)$$

For practical purposes one can think of the error bound as r^{2k}, because $r^{2k}/(1-r^{2k}) \approx r^{2k}$ for r^{2k} small. To have accuracy up to the γth decimal, we let $r = 10^{-\gamma/2k}$.

APPENDIX 6.C: Proof of Lemma 6.1

By Cramer's rule we have that $x_j = |\mathbf{A}_{j+1}|/|\mathbf{A}|$, $j = 0, 1, \ldots, s-1$, where $\mathbf{A}_{j+1}$ is the matrix $\mathbf{A}$ except for the $(j+1)$st column being replaced by $\mathbf{b}$. Since

$|\mathbf{A}| = |\mathbf{A}^T|$, we find

$$|\mathbf{A}| = \begin{vmatrix} s & z_1^s - 1 & \cdots & z_{s-1}^s - 1 \\ s-1 & z_1^s - z_1 & \cdots & z_{s-1}^s - z_{s-1} \\ \vdots & \vdots & \vdots & \vdots \\ 1 & z_1^s - z_1^{s-1} & \cdots & z_{s-1}^s - z_{s-1}^{s-1} \end{vmatrix}$$

$$= \begin{vmatrix} 1 & z_1 - 1 & \cdots & z_{s-1} - 1 \\ 1 & z_1(z_1 - 1) & \cdots & z_{s-1}(z_{s-1} - 1) \\ \vdots & \vdots & \vdots & \vdots \\ 1 & z_1^{s-1}(z_1 - 1) & \cdots & z_{s-1}^{s-1}(z_{s-1} - 1) \end{vmatrix},$$

where the last equality follows by subtracting row $r+1$ from row r for each $r = 1, 2, \ldots, s-1$. Dividing each column $k+1$ by $z_k - 1$ for $k = 1, \ldots, s-1$ yields the following result

$$|\mathbf{A}| = \prod_{k=1}^{s-1}(z_k - 1) \begin{vmatrix} 1 & 1 & \cdots & 1 \\ 1 & z_1 & \cdots & z_{s-1} \\ \vdots & \vdots & \vdots & \vdots \\ 1 & z_1^{s-1} & \cdots & z_{s-1}^{s-1} \end{vmatrix} = \prod_{k=1}^{s-1}(z_k - 1) \prod_{0 \le n < k \le s-1}(z_k - z_n),$$

where the last equality follows from the special form of the determinant of a Vandermonde matrix, see, e.g. Bellman [12].

To compute the determinant of $\mathbf{A}_{j+1}$ we expand this matrix on its $j+1$-st column, which gives $|\mathbf{A}_{j+1}| = (-1)^{j+2}(s - \mu_Y)|\mathbf{B}|$, where

$$|\mathbf{B}| = \begin{vmatrix} s & \cdots & s-j-1 & s-j+1 & \cdots & 1 \\ z_1^s - 1 & \cdots & z_1^s - z_1^{j-1} & z_1^s - z_1^{j+1} & \cdots & z_1^s - z_1^{s-1} \\ \vdots & \vdots & \vdots & \vdots & \vdots & \vdots \\ z_{s-1}^s - 1 & \cdots & z_{s-1}^s - z_{s-1}^{j-1} & z_{s-1}^s - z_{s-1}^{j+1} & \cdots & z_{s-1}^s - z_{s-1}^{s-1} \end{vmatrix}.$$

We then transpose the matrix **B**, and subtract column $k + 1$ from column k to obtain

$$|\mathbf{B}| = |\mathbf{B}^T| = \begin{vmatrix} z_1 - 1 & z_2 - 1 & \cdots & z_{s-1} - 1 \\ z_1(z_1 - 1) & z_2(z_2 - 1) & \cdots & z_{s-1}(z_{s-1} - 1) \\ \vdots & \vdots & \vdots & \vdots \\ z_1^{j-1}(z_1 - 1) & z_2^{j-1}(z_2 - 1) & \cdots & z_{s-1}^{j-1}(z_{s-1} - 1) \\ z_1^{j+1} - z_1^{j-1} & z_2^{j+1} - z_2^{j-1} & \cdots & z_{s-1}^{j+1} - z_{s-1}^{j-1} \\ z_1^{j+1}(z_1 - 1) & z_2^{j+1}(z_2 - 1) & \cdots & z_{s-1}^{j+1}(z_{s-1} - 1) \\ \vdots & \vdots & \vdots & \vdots \\ z_1^{s-1}(z_1 - 1) & z_2^{s-1}(z_2 - 1) & \cdots & z_{s-1}^{s-1}(z_{s-1} - 1) \end{vmatrix}.$$

Dividing each column $k + 1$ by $z_k - 1$ for $k = 1, \ldots, s - 1$ then yields

$$|\mathbf{B}| = \prod_{k=1}^{s-1}(z_k - 1) \begin{vmatrix} 1 & 1 & \cdots & 1 \\ z_1 & z_2 & \cdots & z_{s-1} \\ \vdots & \vdots & \vdots & \vdots \\ z_1^{j-2} & z_2^{j-2} & \cdots & z_{s-1}^{j-2} \\ z_1^{j-1} + z_1^{j} & z_2^{j-1} + z_2^{j} & \cdots & z_{s-1}^{j-1} + z_{s-1}^{j} \\ z_1^{j+1} & z_2^{j+1} & \cdots & z_{s-1}^{j+1} \\ \vdots & \vdots & \vdots & \vdots \\ z_1^{s-1} & z_2^{s-1} & \cdots & z_{s-1}^{s-1} \end{vmatrix}.$$

Since the determinant is a linear operator we can rewrite it as

$$\begin{vmatrix} 1 & 1 & \cdots & 1 \\ z_1 & z_2 & \cdots & z_{s-1} \\ \vdots & \vdots & \vdots & \vdots \\ z_1^{j-2} & z_2^{j-2} & \cdots & z_{s-1}^{j-2} \\ z_1^{j-1} & z_2^{j-1} & \cdots & z_{s-1}^{j-1} \\ z_1^{j+1} & z_2^{j+1} & \cdots & z_{s-1}^{j+1} \\ \vdots & \vdots & \vdots & \vdots \\ z_1^{s-1} & z_2^{s-1} & \cdots & z_{s-1}^{s-1} \end{vmatrix} + \begin{vmatrix} 1 & 1 & \cdots & 1 \\ z_1 & z_2 & \cdots & z_{s-1} \\ \vdots & \vdots & \vdots & \vdots \\ z_1^{j-2} & z_2^{j-2} & \cdots & z_{s-1}^{j-2} \\ z_1^{j} & z_2^{j} & \cdots & z_{s-1}^{j} \\ z_1^{j+1} & z_2^{j+1} & \cdots & z_{s-1}^{j+1} \\ \vdots & \vdots & \vdots & \vdots \\ z_1^{s-1} & z_2^{s-1} & \cdots & z_{s-1}^{s-1} \end{vmatrix}. \qquad (6.\text{C}.1)$$

The matrices in (6.C.1) are very similar to Vandermonde matrices, except that one row has been deleted. As for Vandermonde matrices, the determinants of such matrices have a nice form, see Pólya and Szegö [140], Exercise 10, p. 93, and Neagoe [131]:

$$|\mathbf{B}| = \prod_{k=1}^{s-1}(z_k - 1) \left(\prod_{1 \leq n < k} (z_k - z_n)S_{s-j} + \prod_{1 \leq n < k} (z_k - z_n)S_{s-j-1} \right),$$

where the functions S_{s-j} and S_{s-j-1} are defined as in (6.12). Altogether, this gives that

$$x_j = \frac{|\mathbf{A}_{j+1}|}{|\mathbf{A}|} = (-1)^{j+2}(s - \mu_Y)\frac{S_{s-j} + S_{s-j-1}}{\prod_{k=1}^{s-1}(z_k - 1)},$$

which completes the proof. An alternative proof of Lemma 6.1 has been given in Zhao and Campbell [176]. □

Chapter 7

PERIODIC SCHEDULING

For modelling data transfer organised via a reservation procedure, we proposed in Sect. 2.3 two models, referred to as *fixed boundary model* and *flexible boundary model*. For these models, time is divided into slots, and slots are grouped into frames. The fixed boundary model divides each frame into a fixed number of request and data slots. The flexible boundary model also uses this division, but additionally, the unused data slots (due to lack of data packets) are turned into request slots. For both models, we first consider the queue length at frame boundaries. The stationary queue length distribution in either model can be determined from a rather standard application of the generating function technique (demonstrated in Sect. 6.2 for the discrete bulk service queue). Due to the periodic scheduling, however, it is far less straightforward to analyse the stationary delay. By adopting a technique developed in Bruneel and Kim [28] and Kang and Steyaert [92], we succeed in deriving the probability generating function of the stationary delay.

7.1   Introduction

We elaborated earlier on how the fixed and flexible boundary models arise in the context of data-transmission procedures in cable access networks regulated by a reservation mechanism in which actual data transmission is preceded by a request procedure, see Sect. 2.3. Both the request messages and the actual data transmission take place on the same upstream channel, and hence each slot can either be used for a request message or for a data transmission.

The fixed and flexible boundary models both serve as models for the data queue, defined as those data packets for which transmission has already been requested, but that are still waiting to be transmitted. Clearly, if a slot is used for reservation (request slot), new packets can enter this queue, and if a slot is used for data transmission (data slot), a packet can leave this queue.

Because the transmission delay requires that scheduling decisions are taken in advance, one is naturally led to consider frame-based scheduling. The nature of each slot in the frame is periodically determined and broadcast to all the stations. In this chapter we assume that the timing is such that each station is aware of the layout of a frame before it actually starts.

The chapter is structured as follows. In Sect. 7.2, we describe and motivate the models in more detail. In Sect. 7.3, we derive the pgf of the stationary queue length. A closed-form expression for the pgf of the packet delay for both models is derived in Sect. 7.4. A numerical comparison of the two models is given in Sect. 7.5, followed by some conclusions in Sect. 7.6.

7.2   Model Description

The fixed and flexible boundary models were introduced in Sect. 2.3. We will repeat their exact definitions and refer to Sect. 2.3 for a discussion of the model assumptions in view of a reservation procedure.

Time is assumed to be slotted, with a given slot duration. In case of the fixed boundary model the schedule of each frame is fixed. That is, a frame defined as f consecutive slots consists of c request slots followed by $s := f - c$ data slots. Let the random variable Y_{ni} denote the number of arriving packets during the ith request slot of frame n, and assume that the sequence Y_{ni} is i.i.d. for all n and i. We further assume that packets that arrive during frame n cannot depart from the queue until the beginning of frame $n + 1$. This leads to the recursion

$$X_{n+1} = (X_n - s)^+ + \sum_{i=1}^{c} Y_{ni}, \tag{7.1}$$

where X_n denotes the queue length at the beginning of frame n and $x^+ := \max(0, x)$.

The fixed boundary mechanism seems inefficient, in the sense that if the queue length is smaller than s, it leaves slots unused which could alternatively be scheduled as request slots. This motivates the flexible boundary model in which these unused slots are designated as request slots, yielding the recursion

$$\tilde{X}_{n+1} = (\tilde{X}_n - s)^+ + \sum_{i=1}^{c+(s-\tilde{X}_n)^+} Y_{ni}, \tag{7.2}$$

where, for notational purposes, we add a tilde to the random variables related to the flexible boundary model. We refer to the c request slots that are scheduled at the beginning of every frame as forced request slots.

7.2.1   Scheduling Parameter c

The number of forced request slots c in (7.1) and (7.2) can be interpreted as the amount of bandwidth guaranteed for the request procedure: In each frame there are at least c request slots. For the flexible boundary model there are two, unfortunately conflicting, heuristics that guide a judicious choice of c. On the one hand, setting c small implements a greedy schedule which empties the data queue as quickly as possible, which suggests that this is the appropriate schedule to minimise the data-queue size. On the other hand, setting c large smooths out the arrival process, and intuition suggests that this also helps to reduce the data-queue size. In choosing the right value of c, one should strike the proper balance between these two considerations. One of the goals of this chapter is to investigate the impact of c through a mathematical analysis of the models. Numerical results are presented in Sect. 7.5.

7.3 Queue Length

In this section we derive the pgf of the stationary queue length for both the fixed and flexible boundary model. For each model, we first present the results for the queue length at frame boundaries, from which the results for the queue length throughout a frame follow.

7.3.1 Fixed Boundary Model

Let us denote by Y a random variable that has the same distribution as the number of arriving packets during one request slot, i.e. Y_{ni} are i.i.d. copies of a discrete random variable Y for all n and i. Let $Y(z)$ be the pgf of Y and denote the mean and variance of Y by μ_Y and σ_Y^2, respectively. Clearly, to have stability, it is required that the expected number of arriving packets in a frame is less than the maximum number of packets that can be transmitted in a frame, i.e.

$$c\mu_Y < s. \tag{7.3}$$

We have denoted the queue length at the beginning of frame n by X_n. Then $\{X_n, n \in \mathbb{Z}^+\}$ constitutes a discrete-time Markov chain, with transitions governed by (7.1). As is easily verified, the following conditional expectation holds

$$\mathbb{E}(z^{X_{n+1}}|X_n = k) = \begin{cases} Y(z)^c, & k < s, \\ z^{k-s}Y(z)^c, & k \geq s. \end{cases} \tag{7.4}$$

For reasons of brevity, we introduce the random variable A that is distributed according to the c-fold convolution of the distribution of Y, that is, the pgf of A is given by $A(z) = Y(z)^c$. We denote the mean and variance of A by μ_A and σ_A^2.

Let X be a random variable distributed as the stationary distribution of the queue length, with

$$x_k = \mathbb{P}(X = k) = \lim_{n \to \infty} \mathbb{P}(X_n = k), \quad k = 0, 1, 2, \dots$$

From (7.4) it follows that the pgf of X is given by

$$X(z) = \frac{A(z) \sum_{k=0}^{s-1} x_k (z^s - z^k)}{z^s - A(z)}; \tag{7.5}$$

see (6.7) for the pgf of the stationary queue length in the discrete bulk service queue. In this expression there are still s unknowns $x_0, \ldots, x_{s-1}$, which can be found using the classical approach discussed in Sect. 6.2 of this monograph. In Theorem 6.1 it has been proven that $z^s = A(z)$ has s roots on or within the unit circle. Since a pgf is analytic and well-defined in $|z| \leq 1$, the numerator of $X(z)$ should vanish at each of the roots. This gives s equations. One of the roots equals 1, and leads to a trivial equation. However, the normalisation condition $X(1) = 1$ provides an additional equation. Using l'Hôpital's rule, this condition is found to be

$$s - \mu_A = \sum_{k=0}^{s-1} x_k (s - k), \tag{7.6}$$

which equates two expressions for the mean number of unused data slots per frame. In case some of the roots have a multiplicity larger than one, still a set of linear equations can be constructed that yields the unique solution $x_0, x_1, \ldots, x_{s-1}$, see Remark 6.1 on p. 122.

Explicit expressions for the moments of the queue length can be obtained by taking derivatives of $X(z)$. For example, evaluating the first derivative of $X(z)$ at $z = 1$ yields

$$\mathbb{E}(X) = \frac{\sigma_A^2}{2(s - \mu_A)} + \frac{s + \mu_A}{2} - \sum_{k=0}^{s-1} \frac{x_k (s - k)^2}{2(s - \mu_A)}. \tag{7.7}$$

So far we looked at the queue length at the beginning of a frame. We can also model the behaviour of the queue length throughout a frame. Denote by $X_{[m]}$, $m = 1, 2, \ldots, f$, the steady-state queue length at the end of the mth slot of a frame. The first c slots of a frame are request slots. This implies that the pgf of $X_{[m]}$ is given by

$$X_{[m]}(z) = X(z) Y(z)^m, \quad m = 1, \ldots, c. \tag{7.8}$$

The remaining s slots are data slots, yielding $(m = 1, 2, \ldots, s)$

$$\mathbb{E}(z^{X_{[c+m]}} | X = k) = \begin{cases} A(z), & k < m, \\ A(z) z^{k-m}, & k \geq m. \end{cases} \tag{7.9}$$

Summing over all possible values of X then gives

$$X_{[c+m]}(z) = A(z) \left(\sum_{k=0}^{m-1} x_k + \frac{1}{z^m} \left(X(z) - \sum_{k=0}^{m-1} x_k z^k \right) \right), \quad m = 1, \ldots, s. \tag{7.10}$$

The expectation of the stationary queue length throughout a frame then follows from evaluating the first derivative of (7.8) and (7.10) at $z = 1$. That is, $\mathbb{E}X_{[m]} = \mathbb{E}X + m\mu_Y$ for $m = 1, \ldots, c$, and

$$\mathbb{E}(X_{[c+m]}) = \mathbb{E}(X) + \mu_A - m + \sum_{k=0}^{m-1} x_k(m-k), \tag{7.11}$$

for $m = 1, \ldots, s$. Observe that $\mathbb{E}X_{[f]}$ equals $\mathbb{E}X$ due to the normalisation condition (7.6).

7.3.2 Flexible Boundary Model

For the flexible boundary mechanism, unused data slots are turned into request slots. So, within a frame, the c forced request slots are scheduled first, then the data slots (if any), and finally the additional request slots (if any). The stability condition (7.3) still applies and is equivalent to requiring c to be smaller than $f/(\mu_Y + 1)$.

With $\tilde{X}_n$ representing the queue length at the beginning of frame n, $\{\tilde{X}_n, n \in \mathbb{Z}^+\}$ constitutes a discrete-time Markov chain, with transitions governed by (7.2). Note that the following conditional expectation holds

$$\mathbb{E}(z^{\tilde{X}_{n+1}} | \tilde{X}_n = k) = \begin{cases} Y(z)^{f-k}, & k < s, \\ z^{k-s} A(z), & k \geq s. \end{cases} \tag{7.12}$$

Because in the flexible boundary model all slots are used, the mean number of request slots per frame, denoted by c^*, is fixed and independent of c, i.e.

$$c^* = \frac{f}{\mu_Y + 1}, \tag{7.13}$$

as each request slot requires $1 + \mu_Y$ slots in total: The request slot itself and μ_Y slots for transmitting the packets.

Let $\tilde{X}$ denote a random variable distributed as the stationary queue length distribution, with

$$\tilde{x}_k = \mathbb{P}(\tilde{X} = k) = \lim_{n \to \infty} \mathbb{P}(\tilde{X}_n = k), \quad k = 0, 1, 2, \ldots.$$

From (7.12), it follows that the pgf of $\tilde{X}$ is given by

$$\tilde{X}(z) = \frac{A(z) \sum_{k=0}^{s-1} \tilde{x}_k(z^s Y(z)^{s-k} - z^k)}{z^s - A(z)}. \tag{7.14}$$

As in Sect. 7.3.1, the s roots of $z^s = A(z)$ on or within the unit circle can be used to determine $\tilde{x}_0, \ldots, \tilde{x}_{s-1}$. Using l'Hôpital's rule, the normalisation condition $\tilde{X}(1) = 1$ reads

$$s - \mu_A = \sum_{k=0}^{s-1} \tilde{x}_k (s-k)(\mu_Y + 1), \tag{7.15}$$

which equates two expressions for the mean number of slots per frame that are used for arrivals and departures of packets that arrived in other than the c forced request slots.

The mean queue length in case of the flexible boundary model is given by

$$\begin{aligned}
\mathbb{E}(\tilde{X}) &= \frac{\sigma_A^2}{2(s - \mu_A)} + \frac{\sigma_Y^2}{2(\mu_Y + 1)} + \frac{s + \mu_A}{2} \\
&\quad -(1 - \mu_Y) \sum_{k=0}^{s-1} \frac{\tilde{x}_k (s-k)^2 (1 + \mu_Y)}{2(s - \mu_A)}.
\end{aligned} \tag{7.16}$$

Using the same notation as for the fixed boundary model, the behaviour of the queue length throughout a frame follows from

$$\tilde{X}_{[m]}(z) = \tilde{X}(z) Y(z)^m, \quad m = 1, \ldots, c, \tag{7.17}$$

and, for $m = 1, 2, \ldots, s$,

$$\mathbb{E}(z^{\tilde{X}_{[c+m]}} \mid \tilde{X} = k) = \begin{cases} Y(z)^{c+m-k}, & k < m, \\ A(z) z^{k-m}, & k \geq m, \end{cases} \tag{7.18}$$

and consequently, for $m = 1, 2, \ldots, s$,

$$\tilde{X}_{[c+m]}(z) = A(z) \left(\sum_{k=0}^{m-1} \tilde{x}_k Y(z)^{m-k} + \frac{1}{z^m} \left(\tilde{X}(z) - \sum_{k=0}^{m-1} \tilde{x}_k z^k \right) \right). \tag{7.19}$$

Hence,

$$\mathbb{E}(\tilde{X}_{[m]}) = \begin{cases} \mathbb{E}(\tilde{X}) + m\mu_Y, & m = 1, \ldots, c, \\ (1 + \mu_Y) \sum_{k=0}^{m-c-1} \tilde{x}_k (m - c - k) \\ \quad + c\mu_Y + \mathbb{E}(\tilde{X}) - m + c, & m = c+1, \ldots, f. \end{cases} \tag{7.20}$$

Observe that $\mathbb{E}\tilde{X}_{[f]}$ equals $\mathbb{E}\tilde{X}$ due to the normalisation condition (7.15).

EXAMPLE 7.1 Consider a frame length of 18 slots, and Y distributed according to a Poisson or geometric distribution

$$\mathbb{P}(Y = k) = e^{-\lambda} \frac{\lambda^k}{k!}; \quad \mathbb{P}(Y = k) = (1-p)p^k, \quad k = 0, 1, \ldots,$$

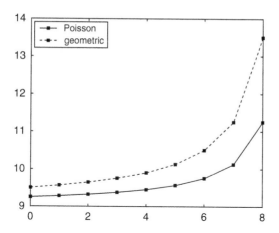

Fig. 7.1. $\mathbb{E}\tilde{X}$ for $f = 18$, $\mu_Y = 1$ for Poisson and geometric distribution, and $c = 0, 1, \ldots, 8$

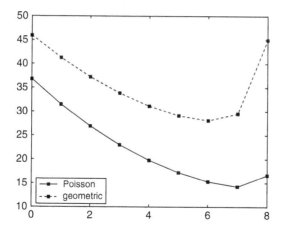

Fig. 7.2. $\sigma_{\tilde{X}}^2$ for $f = 18$, $\mu_Y = 1$ for Poisson and geometric distribution, and $c = 0, 1, \ldots, 8$

both with mean 1 ($\lambda = 1$, $p = 1/2$). The mean and variance of $\tilde{X}$ that correspond to these distributions are shown in Figs. 7.1 and 7.2 for increasing c.

In terms of the mean queue length, having forced request slots at the beginning of the frame is disadvantageous. However, the variance of the queue length is reduced by increasing c, except for high values of c.

REMARK 7.1 In Jacquet et al. [85] a scheduling strategy called *implicit framing* is studied. For this strategy, no frame structure is used and priority is given to data slots. Periods of consecutive request slots are implicitly closed by the first data packet to be transmitted. When all data packets have been transmitted, a new period of consecutive request slots, in which reservation takes place, is restarted. Note that such an implicit framing strategy yields in fact the flexible boundary model with $f = 1$ and $c = 0$. Jacquet et al. [85] demonstrate that within the framework of the flexible boundary model, implicit framing minimises the average delay. In the case of implicit framing, the pgf of $\tilde{X}$ reduces to

$$\tilde{X}(z) = \frac{\tilde{x}_0(zY(z) - 1)}{z - 1} = \frac{1 - zY(z)}{(1 - z)(\mu_Y + 1)}, \tag{7.21}$$

where $\tilde{x}_0$ equals $1/(\mu_Y + 1)$ according to the normalisation condition (7.15). Note that $\tilde{X}$ can be interpreted as the residual lifetime of the random variable $Y + 1$. To see this, divide the time axis in cycles of one request slot plus the number of transmission slots Y granted during that request slot. The residual lifetime is an arbitrary point in a cycle, and since in every slot during this cycle exactly one packet is transmitted, the residual lifetime equals the queue length. We stress, though, that implicit framing is less useful to model the data queue in cable networks, because the transmission delay prevents that a request made by a station in slot s is granted in slot $s + 1$.

7.4   Packet Delay

In deriving the packet delay distribution, the periodic scheduling causes some difficulties, as shown next. We first present a basic result that holds for both the fixed and flexible boundary model, after which we complete the analysis for both models separately.

Assume that the packets are transmitted in order of arrival. Tag an arbitrary packet, and let the random variable T denote the slot within the frame in which this packet arrives, $T \in \{1, 2, \ldots, f\}$. Assume that the packet arrives during slot m, i.e. $T = m$. Introduce $U_{[m]}$ as the number of packets present at the end of the frame that contribute to the tagged packet's delay. Then $U_{[m]}$ consists of the queue length at the end of the frame that was already present at the end of the previous frame, the packets that arrive in the same frame in request slots before T, plus the packets that arrive within the same request slot but before the tagged packet. We then express $U_{[m]}$ in terms of two integer random variables $F_{[m]}$ and $R_{[m]}$

$$U_{[m]} = sF_{[m]} + R_{[m]}, \quad F_{[m]} \geq 0, \quad 0 \leq R_{[m]} \leq s - 1, \tag{7.22}$$

where $F_{[m]}$ denotes the number of complete frames included in the tagged packet's delay, and $R_{[m]}$ the number of packets that will be transmitted during the same frame as the tagged packet, but before it. Introduce $D_{[m]}$ as the random variable representing the delay of a packet that arrives during arrival slot m, defined as

$$D_{[m]} = f - m + fF_{[m]} + c + R_{[m]} + 1. \tag{7.23}$$

That is, $f - m$ slots until the beginning of the next frame, $F_{[m]}$ frames, c forced request slots, $R_{[m]}$ slots within the frame of transmission, and the actual transmission slot of the tagged packet. The pgf of $D_{[m]}$ then reads

$$D_{[m]}(z) = \sum_{i=0}^{\infty} \mathbb{P}(D_{[m]} = i) z^i$$

$$= z^{f-m+c+1} \sum_{j=0}^{\infty} \sum_{k=0}^{s-1} \mathbb{P}(F_{[m]} = j, R_{[m]} = k) z^{fj+k}$$

$$= z^{f-m+c+1} \sum_{j=0}^{\infty} \sum_{k=0}^{s-1} \mathbb{P}(U_{[m]} = sj + k) z^{fj+k}. \tag{7.24}$$

From (7.24) it follows that

$$D_{[m]}(z^s) = z^{s(f-m+c+1)} \sum_{k=0}^{s-1} z^{sk} \vartheta_{mk}(z), \tag{7.25}$$

where the functions $\vartheta_{mk}(z)$ are defined as

$$\vartheta_{mk}(z) = \sum_{j=0}^{\infty} z^{sfj} \mathbb{P}(U_{[m]} = sj + k). \tag{7.26}$$

The problem now is that (7.26) cannot be formulated directly in terms of the pgf of $U_{[m]}$. To achieve this, we use a basic approach that can be found in, e.g. Bruneel and Kim [28] or Kang and Steyaert [92].

7.4.1 Basic Approach

Substituting $l = sj + k$ in (7.26) yields

$$\vartheta_{mk}(z) = \sum_{l=0}^{\infty} z^{(l-k)f} \mathbb{P}(U_{[m]} = l) \sum_{j=-\infty}^{\infty} \delta(l - sj - k), \tag{7.27}$$

with $\delta(n)$ the Kronecker delta function, which equals 1 for $n = 0$ and 0 for all other n. Now invoke the following property

PROPERTY 7.1 For any two integers k and s,

$$\frac{1}{s} \sum_{t=0}^{s-1} a^{tk} = \sum_{j=-\infty}^{\infty} \delta(k - js),$$

where $a = \exp(2\pi i / s)$, $i = \sqrt{-1}$.

Using Property 7.1 we obtain

$$
\begin{aligned}
\vartheta_{mk}(z) &= \sum_{l=0}^{\infty} z^{(l-k)f} \mathbb{P}(U_{[m]} = l) \frac{1}{s} \sum_{t=0}^{s-1} a^{t(l-k)} \\
&= \frac{z^{-kf}}{s} \sum_{t=0}^{s-1} a^{-tk} \sum_{l=0}^{\infty} \mathbb{P}(U_{[m]} = l) z^{fl} a^{tl} \\
&= \frac{z^{-kf}}{s} \sum_{t=0}^{s-1} a^{-tk} U_{[m]}(a^t z^f).
\end{aligned}
\tag{7.28}
$$

Substituting (7.28) into (7.25) yields

$$
\begin{aligned}
D_{[m]}(z^s) &= z^{s(f-m+c+1)} \sum_{k=0}^{s-1} z^{sk} \frac{z^{-kf}}{s} \sum_{t=0}^{s-1} a^{-tk} U_{[m]}(a^t z^f) \\
&= \frac{z^{s(f-m+c+1)}}{s} \sum_{t=0}^{s-1} U_{[m]}(a^t z^f) \sum_{k=0}^{s-1} (z^{-c} a^{-t})^k \\
&= \frac{z^{s(f-m+c+1)}}{s} \sum_{t=0}^{s-1} U_{[m]}(a^t z^f) \frac{1 - (z^{-c} a^{-t})^s}{1 - z^{-c} a^{-t}}.
\end{aligned}
\tag{7.29}
$$

Expression (7.29) gives an explicit formula for the pgf of the packet delay once the pgf of $U_{[m]}$ is known. This leaves us to specify the latter, for which we give separate derivations for the fixed and flexible boundary models.

7.4.2 Packet Delay in Fixed Boundary Model

Let D denote the packet delay for an arbitrary packet. Let Z_0 denote the number of packets at the end of a frame that were already present the frame before, and Z_1 the number of packets within the tagged packet's arrival slot arriving before it. The pgf's of Z_0 and Z_1 are given by

$$
Z_0(z) = \frac{1}{z^s} \left(X(z) + \sum_{k=0}^{s-1} x_k(z^s - z^k) \right),
\tag{7.30}
$$

and

$$
Z_1(z) = \frac{1 - Y(z)}{(1 - z)\mu_Y}.
\tag{7.31}
$$

The pgf of $U_{[m]}$ is then simply given by

$$
U_{[m]}(z) = Z_0(z) Y(z)^{m-1} Z_1(z), \quad m = 1, \dots, c.
\tag{7.32}
$$

Since $\mathbb{P}(T = m) = 1/c$ for $m = 1, \ldots, c$, we have that

$$D(z^s) = \frac{1}{c} \sum_{m=1}^{c} D_{[m]}(z^s),$$

(7.33)

which, combined with (7.29) and (7.32), yields the following result:

THEOREM 7.1 *The pgf of the stationary packet delay in the fixed boundary model* (7.1) *is given by*

$$D(z^s) = \frac{1}{sc} \sum_{t=0}^{s-1} \frac{1 - (a^t z^c)^{-s}}{1 - (a^t z^c)^{-1}} \left\{ z^{s(f+1)} Z_0(a^t z^f) Z_1(a^t z^f) \frac{z^{sc} - A(a^t z^f)}{z^s - Y(a^t z^f)} \right\},$$

(7.34)

where $a = \exp(2\pi i/s)$, $i = \sqrt{-1}$, and $Z_0(z)$ and $Z_1(z)$ as given in (7.30) and (7.31).

The mean packet delay follows from

$$\mathbb{E}(D) = \left[\frac{1}{s} \frac{d}{dz} D(z^s) \right]_{z=1},$$

(7.35)

which gives after tedious but straightforward calculations

$$\mathbb{E}(D) = f + \frac{f\sigma_A^2}{2\mu_A(s - \mu_A)} + \frac{1 + \mu_Y}{\mu_Y} \left(\frac{s}{2} - \sum_{k=0}^{s-1} \frac{x_k(s - k)^2}{2(s - \mu_A)} \right).$$

(7.36)

REMARK 7.2 Alternatively, the mean delay can be derived using Little's law. The queue length at the beginning of an arbitrary slot is given by

$$\frac{1}{f} \sum_{m=1}^{f} \mathbb{E}(X_{[m]}),$$

(7.37)

where $\mathbb{E}(X_{[m]})$ is given by (7.11). The average arrival rate of packets per slot equals $c\mu_Y/f$. Dividing (7.37) by this rate then yields (7.36).

7.4.3 Packet Delay in Flexible Boundary Model

For the flexible boundary model, the derivation of $\tilde{U}_{[m]}(z)$ is somewhat more involved, since all slots within a frame are potential request slots. We first consider the case that $c \geq 1$, while $c = 0$ is covered at the end of this section. Distinguish two events: (a) the tagged packet arrives in one of the forced request slots, and (b) the tagged packet arrives in one of the additional request slots.

Event (a) provides no extra information about the queue length at the beginning of a frame, since the c forced request slots are scheduled every frame. Thus

$$\tilde{U}_{[m]}(z) = \tilde{Z}_0(z)Y(z)^{m-1}Z_1(z), \quad m = 1, \ldots, c, \tag{7.38}$$

where

$$\tilde{Z}_0(z) = \frac{1}{z^s}\left(\tilde{X}(z) + \sum_{k=0}^{s-1}\tilde{x}_k(z^s - z^k)\right). \tag{7.39}$$

Event (b) does provide extra information about the queue length at the beginning of the frame. We know that $\tilde{Z}_0$ equals zero, otherwise there would be no extra request slots. Further, consider the case that the tagged packet arrives in slot $c+1$. This implies that $\tilde{X}$ equals zero. Hence, $\tilde{U}_{[c+1]} = A + Z_1$. Now consider the packet arriving in slot $c+2$. This implies that $\tilde{X}$ equals either zero or one. In the first case it holds that $\tilde{U}_{[c+2]} = A + Y + Z_1$, and in the latter case $\tilde{U}_{[c+2]} = A + Z_1$. Similar reasoning leads to the following expression

$$\tilde{U}_{[m]}(z) = A(z)Z_1(z)\frac{\sum_{k=0}^{m-c-1}\tilde{x}_k Y(z)^{m-c-1-k}}{\sum_{k=0}^{m-c-1}\tilde{x}_k}, \quad m = c+1, \ldots, f. \tag{7.40}$$

Finally, the distribution of T can be determined as follows. Remember that the extra request slots are scheduled at the end of a frame. If a packet arrives in slot $m \in \{c+1, \ldots, f\}$ of a frame, this particular frame has at least $f - m + c + 1$ request slots, and thus at most a queue length of $m - c - 1$ packets at the beginning of the frame. This gives (recall that c^* is the mean number of request slots per frame)

$$\mathbb{P}(T = m) = \begin{cases} \frac{1}{c^*}, & m = 1, \ldots, c, \\ \frac{1}{c^*}\sum_{k=0}^{m-c-1}\tilde{x}_k, & m = c+1, \ldots, f. \end{cases} \tag{7.41}$$

Combining (7.29), (7.38) and (7.40), and conditioning on the request slot distribution given by (7.41) yields an explicit expression for the pgf of the packet delay. We have

$$\begin{aligned}
\tilde{D}(z^s) &= \sum_{m=1}^{f}\mathbb{P}(T = m)\tilde{D}_{[m]}(z^s), \\
&= \sum_{m=1}^{c}\frac{1}{c^*}\tilde{D}_{[m]}(z^s) + \sum_{m=c+1}^{f}\frac{1}{c^*}\sum_{k=0}^{m-c-1}\tilde{x}_k\tilde{D}_{[m]}(z^s), \tag{7.42}
\end{aligned}$$

where $\tilde{D}_{[m]}(z^s)$ is defined as $D_{[m]}(z^s)$ in (7.29), except with $U_{[m]}(a^tz^f)$ replaced by $\tilde{U}_{[m]}(a^tz^f)$. This gives the following result:

THEOREM 7.2 *The pgf of the stationary packet delay in the flexible boundary model* (7.2) *is given by*

$$\tilde{D}(z^s) = \frac{1}{sc^*} \sum_{t=0}^{s-1} \frac{1 - (a^t z^c)^{-s}}{1 - (a^t z^c)^{-1}} \left[z^{s(f+1)} \tilde{Z}_0(a^t z^f) Z_1(a^t z^f) \frac{z^{sc} - A(a^t z^f)}{z^s - Y(a^t z^f)} \right.$$

$$\left. + z^{s(c+1)} A(a^t z^f) Z_1(a^t z^f) \frac{\sum_{k=0}^{s-1} \tilde{x}_k [z^{s(s-k)} - Y(a^t z^f)^{s-k}]}{z^s - Y(a^t z^f)} \right],$$

$$(7.43)$$

where $a = \exp(2\pi i/s)$, $i = \sqrt{-1}$, *and* $\tilde{Z}_0(z)$ *and* $Z_1(z)$ *are given in* (7.39) *and* (7.31).

From (7.42), it follows that

$$\mathbb{E}(\tilde{D}) = \frac{\mu_Y + 1}{\mu_Y} \left[\mathbb{E}(\tilde{X}) + \frac{(s+1)\mu_A - s^2}{2f} + \sum_{k=0}^{s-1} \frac{\tilde{x}_k(s-k)^2(1+\mu_Y)}{2f} \right].$$

$$(7.44)$$

As in case of the fixed boundary model (see Remark 7.2), an alternative derivation of $\mathbb{E}\tilde{D}$ follows from applying Little's law.

In case $c = 0$, the basic approach as described in Sect. 7.4.1 is not needed. It is then straightforward to derive that

$$\tilde{U}_{[m]}(z) = Z_1(z) \frac{\sum_{k=0}^{m-1} \tilde{x}_k Y(z)^{m-1-k}}{\sum_{k=0}^{m-1} \tilde{x}_k}; \qquad \mathbb{P}(\tilde{T} = m) = \frac{1}{c^*} \sum_{k=0}^{m-1} \tilde{x}_k,$$

and that the pgf of the packet delay is given by

$$\tilde{D}_0(z) = \frac{Z_1(z)}{fc^*} \sum_{m=1}^{f} z^{f-m+1} \sum_{k=0}^{m-1} \tilde{x}_k Y(z)^{m-1-k}. \qquad (7.45)$$

EXAMPLE 7.2 The distribution of the packet delay has a typical form. For $f = 9$, Y geometrically distributed with mean 1, Fig. 7.3 displays the packet delay distribution for $c = 0, 2, 4$, where we have used the method described in Appendix 6.B. First note that the minimum delay corresponds to a packet that arrives in the last slot of a frame and is immediately transmitted in the slot $c + 1$ of the next frame. The oscillating effect is due to the frame structure, and becomes stronger for higher values of c.

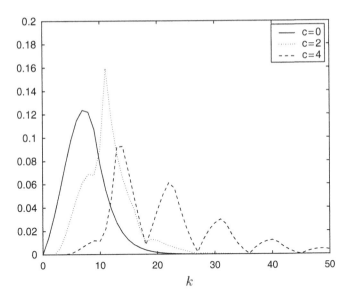

Fig. 7.3. $\mathbb{P}(\tilde{D} = k)$ for $f = 9$, $c = 0, 2, 4$, and Y geometrically distributed with mean 1

7.5   Numerical Results

In this section we first present a numerical comparison between the fixed and flexible boundary models. Next, for the flexible boundary model, we investigate the impact of different values of c on various queue length and delay characteristics.

7.5.1   Fixed vs. Flexible Boundary Model

We assume that the load, defined as the mean number of packets arriving per frame, is the same for the fixed and flexible boundary models, being $c\mu_Y$ and $c^*\mu_Y = f\mu_Y/(1 + \mu_Y)$, respectively. Thus, for a fair comparison, we choose the appropriate values of μ_Y for which the load is the same for both models. For convenience, we further assume that Y is Poisson distributed.

Figures 7.4 and 7.5 display the mean and variance of the packet delay for $f = 9$, $c = 2$ and various load values. For a load of 2, $\mu_Y = 1$ for the fixed boundary model and $\mu_Y = 2/7$ ($c^* = 7$, $c^*\mu_Y = 2$) for the flexible boundary model. In terms of the mean packet delay, the flexible boundary model clearly outperforms its fixed counterpart.

For the flexible boundary model, a low load yields relatively many unused data slots that are used as additional request slots. The variation in request slots per frame inherent to such type of scheduling then causes a higher packet delay variance than in the fixed boundary model.

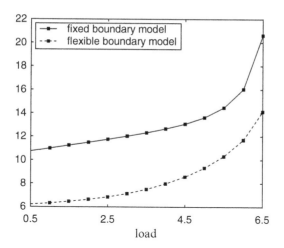

Fig. 7.4. Mean packet delay, fixed vs. flexible boundary model, $f = 9$, $c = 2$, Y Poisson distributed

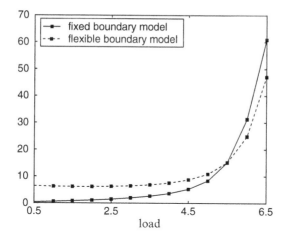

Fig. 7.5. Packet delay variance, fixed vs. flexible boundary model, $f = 9$, $c = 2$, Y Poisson distributed

7.5.2 Influence of c in Flexible Boundary Model

We now investigate the impact of different values of c for the flexible boundary model on various queue length and delay characteristics. Table 7.1 contains queue length characteristics for $f = 9$, $c = 0, 2, 4$, and Y is Poisson or geometrically distributed with mean 1. Table 7.2 contains delay characteristics for the same settings.

As we have seen in Example 7.1, increasing c is disadvantageous in terms of the mean queue length, while the variance of the queue length is often reduced

Table 7.1.   Characteristics of the queue length for $f = 9$ and $\mu_Y = 1$

		$\mathbb{E}\tilde{X}$	$\mathbb{V}\mathrm{ar}\tilde{X}$	$\mathbb{P}(\tilde{X} > 10)$	$\mathbb{P}(\tilde{X} > 20)$	$\mathbb{P}(\tilde{X} > 50)$
Poisson	$c = 0$	4.75	11.75	0.0639	0.0003	0.0000
	$c = 2$	4.95	7.97	0.0408	0.0001	0.0000
	$c = 4$	6.75	10.93	0.1245	0.0019	0.0002
Geometric	$c = 0$	5.00	16.67	0.1042	0.0026	0.0001
	$c = 2$	5.40	14.07	0.0995	0.0020	0.0000
	$c = 4$	9.00	34.63	0.3197	0.0471	0.0064

Table 7.2.   Characteristics of the packet delay for $f = 9$ and $\mu_Y = 1$

		$\mathbb{E}\tilde{D}$	$\mathbb{V}\mathrm{ar}\tilde{D}$	$\mathbb{P}(\tilde{D} > 10)$	$\mathbb{P}(\tilde{D} > 20)$	$\mathbb{P}(\tilde{D} > 30)$
Poisson	$c = 0$	6.92	7.60	0.0926	0.0020	0.0000
	$c = 2$	8.57	8.82	0.5437	0.0039	0.0001
	$c = 4$	13.66	32.03	0.9550	0.2800	0.0327
Geometric	$c = 0$	7.63	11.84	0.1767	0.0028	0.0003
	$c = 2$	11.46	17.10	0.6075	0.0353	0.0014
	$c = 4$	21.40	96.86	0.9568	0.4855	0.1812

due to the stabilising effect on the arrival process. The same can be seen from the results in Table 7.1. Increasing c reduces the flexibility of the system, which is the reason that both the mean and variance of the queue length increase when c gets large, $c = 4$ in this example.

The mean and variance provide only partial information on the underlying distribution function. We therefore consider some tail probabilities. Note that in Table 7.1, the probability that $\tilde{X}$ gets larger than 10 is the smallest for $c = 2$, for both the Poisson and geometric distribution. Depending on the relevant performance characteristic, one can determine the optimal value of c.

For the delay characteristics in Table 7.2 we do not see a stabilising effect. This is mainly due to a convention in our model that packets cannot be scheduled until the next frame. Because of this convention, our definition of delay includes the time between the request slot and the beginning of the next frame. In this way small values of c are favoured: In this case there are few scheduled request slots and many additional request slots. These additional request slots are located near the end of a frame and bring along a smaller delay until the beginning of the next frame than the scheduled request slots that are located at the beginning of a frame.

7.6   Conclusion

We have presented time-slotted queueing models that describe the frame-based scheduling of request and data slots. The fixed boundary model uses a fixed number of request slots per frame, whereas the flexible boundary model designates in addition the unused, due to lack of data packets, data slots as request slots. For both models we presented the pgf of the stationary queue length and the packet delay. For deriving the pgf of the packet delay we used a technique specifically designed to deal with the periodic scheduling. In Van Leeuwaarden [110] this technique has been applied to derive the delay distribution of vehicles in the fixed-cycle traffic-light queue.

For the flexible boundary model we use an optimistic scenario, in the sense that the packets that arrive in the additional request slots can all be transmitted at the beginning of the next frame. If this is not the case, the delayed flexible model, which is the topic of Chap. 8, is more appropriate. The impact of the number of forced request slots, c, on the performance characteristics has been briefly touched upon. This issue will be pursued in much greater depth in Chap. 8.

Chapter 8

RESERVATIONS WITH TRANSMISSION DELAYS

Reservation procedures are often used on communication channels that are characterised by large round-trip times. As a consequence, a successful reservation must be separated in time from the corresponding data transmission. We propose a queueing model for this situation. Moreover, we investigate scheduling strategies: Strategies to divide the bandwidth between the request process and the data-transmission process. In particular, we consider 'static' scheduling and 'adaptive' scheduling.

The static scheduling strategy leads us to formulate a model which is an extension of the bulk service queueing models considered in Chaps. 6 and 7. Its stationary distribution can be obtained by means of classical techniques, but this relies heavily on numerical methods. However, combining a method of Kingman, bounding techniques, and a heuristic argument, we are able to derive approximating bounds for the mean queue length. The bounds are complemented by simulations to deduce an interesting non-monotonicity property of the mean queue length in this model. This result also suggests how to adapt the two scheduling strategies. Additionally, we compare the two strategies and show that they can be used to improve the organisation of the reservation process.

8.1   Introduction

In Chap. 7, we introduced the fixed and the flexible boundary models as queueing models for reservation procedures. However, the large round-trip times, that are often inherent to the communication channel, may complicate the organisation of the reservation procedure. This typically happens in cable networks, as extensively reviewed in Sect. 1.3.

As discussed in Sect. 2.3, this also hampers the applicability of both the fixed and flexible boundary model. To amend this situation, we are led to consider the following recursion for the data queue:

$$X_{n+1} = (X_n - f + c_n)^+ + \sum_{i=1}^{c_{n-d}} Y_{n-d,i}, \qquad (8.1)$$

with $x^+ := \max(0, x)$ and

$$Y_{ni} \text{ i.i.d.}, \quad Y_{ni} \overset{d}{=} Y. \qquad (8.2)$$

In (8.1), X_n denotes the size of the data queue at the start of frame n and $c_n \leq f$ is the number of request slots in frame n, so that $f - c_n$ slots can be used to serve the data queue. The actual data transmissions can be scheduled $d + 1$ frames after the successful request. The recursion with $d = 1$ or $d = 2$ is relevant for cable networks. The Y_{ni} describe the request sizes: $Y_{ni} = 0$ if the request was not successful and Y_{ni} equals the request size otherwise. Their assumed independence can be defended on the basis of the properties of the request process.

The recursion (8.1) is incomplete, as it does not fix the properties of the c_n. As c_n is the number of request slots in frame n, specification of c_n requires a scheduling policy to divide the bandwidth between the request and the data-transmission processes. This policy should be based on what is currently known to the scheduler: The data-queue sizes $\{X_l : l \leq n\}$ and the recent history of the scheduling process $\{c_l, l \leq n - 1\}$. Of course, one expects the current size of the data queue, X_n, and the recent history of the scheduling process, $\{c_l, n - d \leq l \leq n - 1\}$, to be sufficient for an optimal policy. Additionally, we require that $c_n \geq f - X_n$ so that it is guaranteed that slots are always used for either requests or data transmissions, i.e. no slots are unused.

Below we consider the scheduling in more detail. Note that there are, under the scheduling constraints, on average $f/(1 + \mathbb{E}(Y))$ request slots per frame: As one slot is used for the request itself and $\mathbb{E}(Y)$ slots are needed, on average, for the corresponding data transmission. Hence the average traffic intensity equals $\Lambda := f\mathbb{E}(Y)/(1+\mathbb{E}(Y))$ packets per frame for each scheduling strategy satisfying our constraints.

8.1.1 Static and Adaptive Scheduling

The c_n in (8.1) can be interpreted as the amount of bandwidth for the requests: In each frame there are $c_n \leq f$ time slots used for requests. The remaining slots are used for data transmission and let the data queue decrease by $f - c_n$.

As in the no delay case considered in Chap. 7, there are two, unfortunately conflicting, heuristics that guide a judicious scheduling policy. On the one hand, setting $c_n = (f - X_n)^+$ implements a greedy schedule. Indeed, this schedule gives priority to data transmissions and schedules request slots only in case there is no more data to transmit. This empties the data queue as quickly as possible. On the other hand, setting c_n larger guarantees some bandwidth to the request process, which smoothens the arrival process.

In choosing the right policy for c_n, one should strike the proper balance between these two considerations. Not surprisingly, the choice of policy for c_n has received quite some attention in the literature, see [73, 74, 78, 150], as reviewed in Sect. 1.4. The scheduling policies proposed in these references were motivated by ad-hoc arguments and their evaluation was by means of simulation.

A common strategy to guarantee bandwidth for the requests is to choose

$$c_n = c + (f - c - X_n)^+, \tag{8.3}$$

which guarantees c slots per frame to the request process, and gives priority to data transmissions in the remainder of the time slots. It constitutes a natural generalisation of the greedy policy, which sets c to zero, and leads to the recursion

$$X_{n+1} = (X_n - f + c)^+ + \sum_{i=1}^{c+(f-c-X_{n-d})^+} Y_{n-d,i}. \tag{8.4}$$

The queueing model defined via (8.4) is referred to as the delayed bulk service queue. It is named the acquisition queue in [53], and generalises the models considered in Chap. 7. The specification of a scheduling policy now boils down to an appropriate choice for c. It is one of the goals of this chapter to get a better understanding of this scheduling policy through a mathematical analysis. In the remainder, we call this scheduling policy the *static* policy as it does not adapt the choice of the c_n to the recent history of the scheduling process.

We also consider a class of *adaptive* scheduling strategies in which c_n can be chosen in each frame, depending on the recent history of both data queue and scheduling process. We derive an efficient scheduling strategy for choosing the value c_n based on the queue length at the beginning of the frame and the number of request slots scheduled in the previous d frames.

8.1.2 Outline of this Chapter

In Sect. 8.2, we consider the delayed bulk service queue defined in (8.4). The case with $d = 0$ is considered in Chap. 7 and can be solved by means of classical techniques: In this case, the recursion defines a one-dimensional Markov chain, and the probability generating function of the stationary distribution of the data-queue length can be obtained. For $d \geq 1$, the recursion defines a $d + 1$-dimensional Markov chain. Remarkably, the approach from Chap. 7 can be adapted, and the pgf of the stationary distribution of the queue size can be expressed in terms of the roots of some equation. This is shown in Appendix 8.B. This appendix is an excerpt from Denteneer et al. [53], to which we refer for more details and examples.

However, this solution relies heavily on numerical methods and does not provide direct insight, nor does it readily lead to scheduling algorithms. Therefore, we will concentrate on an analysis of (8.4) partly based on heuristic arguments. In particular, we derive approximating bounds for the mean data-queue length under the stationary distribution. In Sect. 8.2.2, we exploit a method of Kingman [95] to express the mean queue length in terms of moments of the arrival distribution, a term related to the idle time, and a correlation term. We further present bounds for the term related to the idle time, using methods from Denteneer et al. [50]. These rigorous results are complemented, in Sect. 8.2.2, with a heuristic derivation to approximate the correlation term. The bounds and the approximation together yield approximations for the mean stationary queue length.

The approximation suggests a remarkable non-monotonicity property of the delayed bulk service queue: The mean queue length is not monotonically increasing in the average traffic intensity. This property is verified by simulation in Sect. 8.2.3, where we also verify the accuracy of the approximation.

Then we turn to scheduling. In Sect. 8.3, we first consider (8.4) and static scheduling. Then we turn to (8.1) and adaptive scheduling, for which we utilise the analysis of the delayed bulk service queue in the following way. The analysis reveals that the non-monotonicity of the delayed bulk service queue is due to the autocorrelation of the arrival process: Increasing the traffic intensity decorrelates the arrival process and this causes the expected queue length to decrease. This phenomenon points out an adaptive schedule to choose c_n. In Sect. 8.4, the two schedules are compared by means of simulation. We end this chapter with a summary and some conclusions.

8.2   The Delayed Bulk Service Queue

We consider the delayed bulk service queue, defined via (8.4). Sections 8.2.1 and 8.2.2 are devoted to an analysis of the mean stationary queue length. In Sect. 8.2.3 we study the non-monotonicity property.

8.2.1   Analysis

We denote the mean and variance of the random variable Y by μ_Y (or $\mathbb{E}Y$) and σ_Y^2. We use the shorthand $s := f - c$. In the sequel, we assume that $c\mu_Y < f - c$, so that the Markov Chain, defined via Recursion (8.4) is ergodic, see Lemma 8-8.B.1 in Appendix 8.B. There we will also describe a method to derive the stationary distribution of (8.4).

To get direct insight in the mean queue length for general $d \geq 0$, we apply a method used by Kingman [95]. This method is based on the manipulation of

$$M_n = (s - X_n)^+, \quad P_n = (X_n - s)^+. \tag{8.5}$$

For these variables, the following obvious relations hold:

$$X_n - s = P_n - M_n, \quad (X_n - s)^2 = P_n^2 + M_n^2. \tag{8.6}$$

We will use X^d to denote a random variable distributed according to the stationary distribution of the queue length process as defined by (8.4), and M^d to denote a random variable that follows the same distribution as $(s - X^d)^+$. We then have that

THEOREM 8.1 *The mean queue length in the delayed bulk service queue with delay parameter d is given by*

$$\mathbb{E}(X^d) = \frac{c\sigma_Y^2}{2(s - c\mu_Y)} + \frac{\sigma_Y^2}{2(1 + \mu_Y)} + \frac{s + c\mu_Y}{2}$$

$$+ \mathbb{E}((M^d)^2)\frac{\mu_Y^2 - 1}{2(s - c\mu_Y)} + \mathbb{E}(R^d)\frac{\mu_Y}{s - c\mu_Y}, \tag{8.7}$$

where

$$\mathbb{E}(R^d) = \lim_{n \to \infty} \mathbb{E}(P_n M_{n-d}). \tag{8.8}$$

Proof See Appendix 8.A. □

Expression (8.7) for $\mathbb{E}(X^d)$ contains two unknown terms: A term $\mathbb{E}((M^d)^2)$ related to the idle time and a correlation term $\mathbb{E}(R^d)$. Now

$$\mathbb{E}((M^d)^2) = \sum_{j=0}^{s-1} \mathbb{P}(X^d = j)(s - j)^2 \tag{8.9}$$

can be satisfactorily bounded in the following way. Since

$$\left(\sum_{j=0}^{s-1} \mathbb{P}(X^d = j)(s - j)\right)^2 \leq \sum_{j=0}^{s-1} \mathbb{P}(X^d = j)(s - j)^2 \leq s\sum_{j=0}^{s-1} \mathbb{P}(X^d = j)(s - j), \tag{8.10}$$

and

$$\sum_{j=0}^{s-1} \mathbb{P}(X^d = j)(s - j) = \frac{s - c\mu_Y}{1 + \mu_Y}, \tag{8.11}$$

we have

$$\left(\frac{s - c\mu_Y}{1 + \mu_Y}\right)^2 \leq \mathbb{E}((M^d)^2) \leq s\frac{s - c\mu_Y}{1 + \mu_Y}. \tag{8.12}$$

These bounds, and some further improvements, have been presented in [50].

In case $d = 0$, we obviously have that $R^d = 0$. Then, combining (8.7) with (8.12) yields bounds for the mean queue length. The bounds are sharp for the heavy-traffic case in which $c\mu_Y \to s$. They have an advantage over exact procedures to calculate $\mathbb{E}(X^0)$ in that they depend on the distribution of Y only via the first two moments. Hence, these bounds are appropriate for usage in case the traffic model is only partly specified; see [51] for an example. Moreover, they are simple explicit expressions and readily amenable to back-of-the-envelope calculations, and this can give an advantage over exact procedures that usually involve numerical calculations. For $d \geq 1$, no simple bounds on the correlation term are known; see [53] for an exact, but involved, expression. Instead, we derive an approximation for $\mathbb{E}(R^d)$ in the next section.

8.2.2   An Approximation for the Correlation Term

In this section we use a heuristic argument to construct an approximation for $\mathbb{E}(R^d)$, which, together with the bounds (8.12), yields approximations for $\mathbb{E}(X^d)$. The argument is based on the inspection of the sample paths of various realisations of the process defined by (8.4).

One such sample path is shown in Fig. 8.1, where $d = 100$, $c = 0$, and Y geometrically distributed with $\mu_Y = 1.25$. The figure shows the evolution of the queue length at frame boundaries after a long initial warm-up period. The

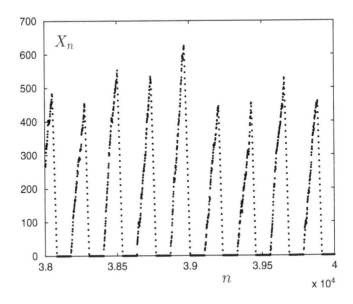

Fig. 8.1. Sample path of the process defined by (8.4), for $f = 18$, $d = 100$, $c = 0$, and Y geometrically distributed with $\mu_Y = 1.25$. The sample path comes from a simulation that started with an empty queue, and the results are displayed for frame 38,000 until frame 40,000

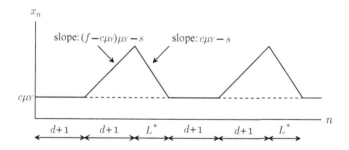

Fig. 8.2. Deterministic approximation x_n of a sample path of (8.4) for $\mu_Y > 1$

sample path has settled on a cyclic pattern. Each cycle can be subdivided into three distinct parts, as illustrated in Fig. 8.2. First, there is an interval of length $d+1$, in which the queue length equals 0. In the second interval, also of length $d+1$, the queue length increases. Finally, in the third interval (the length of this interval is specified below), the data-queue length is drained until it hits zero, upon which a new cycle starts.

We conjecture that this is the typical behaviour of the sample paths in case $\mu_Y > 1$ and $d > 0$, irrespective of the actual distribution of Y. This conjecture suggests that we can construct a deterministic approximation of the sample path. Our heuristic approximation of $\mathbb{E}(R^d)$ is then obtained by evaluating $\mathbb{E}(R^d)$ for this deterministic approximation. More formally, we define the deterministic process x_n via (8.4) with Y_{ni} replaced by its expected value (see Fig. 8.2):

$$x_{n+1} = (x_n - f + c)^+ + \sum_{i=1}^{c+(f-c-x_{n-d})^+} \mu_Y. \qquad (8.13)$$

Given initial values $x_1 = \cdots = x_{d+1} = c\mu_Y$, it is easy to see that (8.13) yields for $j = 1, \ldots, d+1$

$$x_{d+1+j} = j(f - c\mu_Y)\mu_Y - (j-1)(f-c),$$

because in this period those packets join the data queue that were generated in the $f - c\mu_Y$ request slots $d+1$ frames earlier, while packets are transmitted from the queue at maximum rate $f - c$ packets per frame. At the end of this period, the queue has built up to the level $(d+1)(f - c\mu_Y)\mu_Y - d(f-c)$, after which the queue is drained at rate $(f - c) - c\mu_Y$. This yields

$$x_{2(d+1)+j} = (d+1)(f - c\mu_Y)\mu_Y - (d+j)(f-c) + jc\mu_Y,$$

for $j = 1, \ldots, L^*$. Here L^* is the smallest value l for which $x_{2(d+1)+l}$ hits $c\mu_Y$. Consequently, L^* can be calculated from $x_{2(d+1)+L^*} = c\mu_Y$, i.e.

$$L^* = \frac{(d+1)(f - c\mu_Y)\mu_Y - d(f - c)}{f - c - c\mu_Y}. \tag{8.14}$$

After instant $2(d+1) + L^* - 1$ the sequence repeats itself. Hence the cycle length equals $L = 2(d+1) + L^* - 1 = (d+1)(\mu_Y + 1)$. We therefore approximate $\mathbb{E}(R^d)$ as follows

$$
\begin{aligned}
\mathbb{E}(R^d) &\approx \lim_{n \to \infty} \frac{1}{T} \sum_{n=1}^{T} (x_n - s)^- (x_{n+d} - s)^+ \\
&\approx \frac{1}{L} \sum_{n=1}^{L} (x_n - f + c)^- (x_{n+d} - f + c)^+. \tag{8.15}
\end{aligned}
$$

where $x^- = -\min(0, x)$. Now for $\mu_Y > 1$, we can approximate the second sum in (8.14) by the terms $n = 2, \ldots, d+1$, so that

$$
\begin{aligned}
\mathbb{E}(R^d) &\approx \frac{1}{L} \sum_{j=1}^{d} (c\mu_Y - f + c)^- \\
&\qquad \times (j(f - c\mu_Y)\mu_Y - (j-1)(f - c) - f + c)^+ \\
&= \frac{1}{L} \frac{1}{2} d(d+1)(f - c - c\mu_Y)((f - c\mu_Y)\mu_Y - f + c)^+ \\
&= \frac{1}{2} \frac{d}{\mu_Y + 1} (f - c - c\mu_Y)((f - c\mu_Y)\mu_Y - f + c)^+. \tag{8.16}
\end{aligned}
$$

Substituting (8.16) into (8.7) yields the following approximation for $\mathbb{E}(X^d)$:

$$
\begin{aligned}
\mathbb{E}(X^d) &\approx \frac{c\sigma_Y^2}{2(s - c\mu_Y)} + \frac{\sigma_Y^2}{2(1 + \mu_Y)} + \frac{s + c\mu_Y}{2} + \mathbb{E}((M^d)^2)\frac{\mu_Y^2 - 1}{2(s - c\mu_Y)} \\
&\quad + \frac{1}{2} d \frac{\mu_Y}{\mu_Y + 1} ((f - c\mu_Y)\mu_Y - f + c)^+. \tag{8.17}
\end{aligned}
$$

The bounds in (8.12) for $\mathbb{E}((M^d)^2)$ can now be used to obtain explicit expressions.

In order to assess the quality of the approximation we have carried out a number of simulations. We consider Y to be distributed according to a Poisson distribution; simulations with a geometric distribution showed similar results and have been left out. The frame length has been set to $f = 18$, and we varied the traffic intensity Λ, the transmission delay d, and the number of forced request slots c. The mean queue length has been evaluated on an interval of 1,000,000 frames, after an initial warm-up period of 200,000 frames.

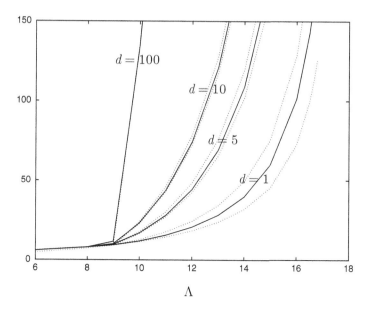

Fig. 8.3. $\mathbb{E}(X^d)$ for Poisson Y, $c = 0$ and various values of d

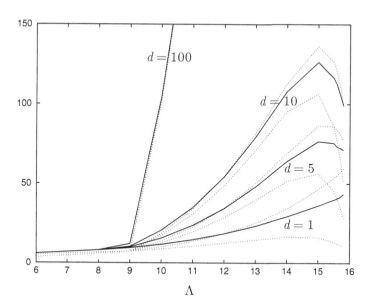

Fig. 8.4. $\mathbb{E}(X^d)$ for Poisson Y, $c = 2$ and various values of d

The performance curves obtained by simulation (solid lines) for $c = 0$ and $c = 2$ are shown for for Poisson Y in Figs. 8.3 and 8.4. Additionally, the approximating lower and upper bounds, obtained by substitution of (8.12) in

(8.17), are shown as dashed lines. These simulations lead us to conclude that the approximations are excellent for $c = 0$.

There are some marked differences between the results displayed in Figs. 8.3 and 8.4. Most importantly, the approximation by (8.17) is less accurate in case $c = 2$ than it is in case $c = 0$. This is so in particular for the higher traffic intensities. Additionally, the approximating bounds are not actual bounds in this case.

From these and many other examples we conclude that (8.17) is in general sharp, but breaks down in heavy traffic conditions for $c > 0$. The latter is because the deterministic approximation to the sample path is less convincing for $c > 0$ and heavy traffic conditions.

8.2.3   A Non-Monotonicity Property

The approximation (8.17) suggests various interesting properties for $\mathbb{E}(X^d)$. Firstly, we consider $\mathbb{E}(X^d)$ as a function of d. The approximation then suggests that d has no impact on the mean queue length in case $\mu_Y \leq 1$. However, in case $\mu_Y > 1$, $\mathbb{E}(X^d)$ increases linearly with d. It follows in particular that the correlation term $\mathbb{E}(R^d)$ is the dominating term in the expression for $\mathbb{E}(X^d)$ and that $\mathbb{E}(X^d)$ grows without bounds for d tending to infinity.

This also suggests that the mean queue length is not necessarily monotonic in the traffic intensity for $d > 0$ and $c > 0$. To see this, observe that the approximation (8.16) of the correlation term $\mathbb{E}(R^d)$ is not monotonic in μ_Y. This follows easily as this approximation is nonnegative, and equals 0 both for the low traffic intensities $\mu_Y \leq 1$ and the maximum traffic intensity $\mu_Y = (f - c)/c$. Now for large values of d, the correlation term will dominate the expression for the mean queue length, which then is non-monotonic too.

In order to verify this non-monotonicity, we carried out a substantial number of simulations, for the the Poisson distribution, with $c = 1$ and the average traffic intensity per frame close to the stability boundary $\Lambda = 17$. The resulting performance curves is presented in Fig. 8.5 along with the approximating bounds based on (8.17). Similar simulations were carried out for the geometric distribution. These simulations show convincingly that the non-monotonicity is a real feature of the model and not a mere artifact of the approximations. These figures also show that this effect can be very substantial.

This non-monotonicity can be explained informally as follows. Observe that the input to the data queue consists of two sources: $(X_n - s)^+$ and a sum which increases in $(s - X_{n-d})^+$. As the traffic intensity approaches the stability bound, the cyclic behaviour of the sample paths vanishes which decorrelates the two input sources to the data queue. Hence, increasing the traffic intensity causes the input to be less bursty, and this results in a smaller mean queue length. Another way to see this is by observing that the bursts following periods

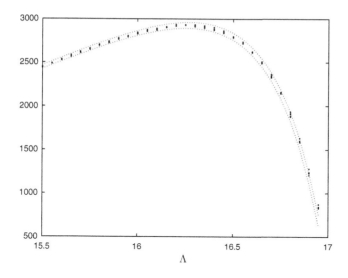

Fig. 8.5. $\mathbb{E}(X^d)$ for Poisson Y, $c = 1$ and $d = 100$. $\mathbb{E}(X^d)$ is not monotonous in the traffic intensity

in which the system is, relatively, empty are caused by an inflow of magnitude $(f - c - c\mu_Y)\mu_Y + c\mu_Y = (f - c\mu_Y)\mu_Y$. Now this latter expression is non-monotonic in μ_Y.

This non-monotonic behaviour, though remarkable, is not uncommon in systems that involve control and feedback delay. Situations in which these characteristics lead to unwanted oscillations and increased delay occur if the traffic dynamics can be expressed via a difference equation or differential equation that involves a delayed response, see, e.g. [65, 76, 91].

8.3 Adaptive Scheduling Strategies

In the previous section, we have analysed the delayed bulk service queue and obtained an approximation (8.17) for the expected queue length. Moreover, we have revealed a remarkable non-monotonicity property of the expected queue length. Our results can be utilised in two ways to construct efficient scheduling strategies to divide the bandwidth between the request and the data-transmission phase.

Firstly, they aid in choosing the parameter c in the delayed bulk service queue: By minimising (8.17) with respect to c we obtain a value for c, adapted to the traffic intensity, which minimises the expected queue length. Of course, this observation must be combined with a method to estimate the traffic intensity in order to arrive at a practical scheduling algorithm. This approach is referred to as static scheduling. The strategy is illustrated in Fig. 8.6, where

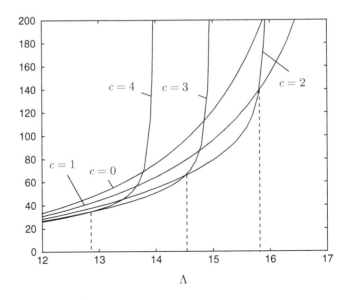

Fig. 8.6. $\mathbb{E}X^d$ for geometric arrivals and $d = 3$

it is shown that c can be chosen adaptively, depending on the traffic intensity, to decrease the expected queue length.

Secondly, these results help in formulating a more general adaptive scheduling strategy: The non-monotonicity was due to the autocorrelation, or burstiness, of the arrival process, and improved, adaptive, scheduling strategies can be constructed by controlling the autocorrelation. This is considered in more detail below.

We have seen in Sects. 8.2.2 and 8.2.3 that the transmission delay results in a cyclic behaviour and a strongly correlated arrival process. This might have severe consequences for the mean queue length (see Expression (8.7)), since the correlation term $\mathbb{E}(R^d)$ becomes dominant in high-load situations. One way to smooth the arrival process is static scheduling as considered above. Here, we consider an alternative method to reduce the correlation of the arrival process. We will do this by introducing a scheduling strategy that does not only vary the number of request slots per frame as a function of the queue length at the beginning of the frame (as for static scheduling), but also allows for the number of request slots in a frame to depend on the number of request slots scheduled in the previous d frames. We have referred to this strategy as adaptive scheduling.

Denote by c_n the number of request slots scheduled in frame n. The evolution equation of the queue length at frame boundaries then becomes (8.1). Let us now recall where the cyclic behaviour observed in Sect. 8.2.2 comes from. The arrival process is coupled to the queue length such that more packets arrive when the queue is small, and less packets arrive when the queue is long. This type of control is expected to lead to a smoother distribution of the number of arriving packets over time. However, the transmission delay upsets the balance. The impact of a corrective decision, like more arrivals if the system is less busy, is only seen d frames later. If the system is busier d frames later, the extra arrivals might have just the opposite effect. This phenomenon of control decisions that have the opposite effect as one had in mind is precisely what is captured by the correlation term:

$$
\begin{aligned}
\mathbb{E}(R^d) &= \lim_{n \to \infty} \mathbb{E}(P_n M_{n-d}) \\
&= \lim_{n \to \infty} \sum_{j=0}^{s-1} \mathbb{P}(M_{n-d} = j) \sum_{k=0}^{\infty} \mathbb{P}(P_n = k | M_{n-d} = j) j k \\
&= \lim_{n \to \infty} \sum_{j=0}^{s-1} \mathbb{P}(X_{n-d} = s - j) \\
&\quad \times \sum_{k=0}^{\infty} \mathbb{P}(X_n = s + k | X_{n-d} = s - j)(s - j)(s + k).
\end{aligned}
$$
(8.18)

So, $\mathbb{E}(R^d)$ might be viewed as a measure for the performance of a scheduling strategy: A high value of $\mathbb{E}(R^d)$ indicates that the scheduling strategy balances the input poorly, and ideally $\mathbb{E}(R^d)$ equals zero. Obviously, it holds that the larger the transmission delay d, the less unlikely it is that the relatively simple static scheduling balances the input well.

Our primary goal is to reduce the mean queue length by choosing an adaptive scheduling strategy that balances the input properly despite a substantial delay. Denote by c^* the mean number of request slots per frame, given by

$$
c^* = \frac{f}{1 + \mu_Y}.
$$
(8.19)

The mean number of arriving data packets per frame, denoted by Λ, is then given by

$$
\Lambda = c^* \mu_Y = \frac{f \mu_Y}{1 + \mu_Y}.
$$
(8.20)

In balancing the input, one would want Λ packets to arrive to the queue per frame. This is not feasible, since we are dealing with a stochastic process, but it might serve as a guiding principle. Say we are at the beginning of frame n. What do we know about the number of arriving packets in the next d frames? We have scheduled $c_{n-d} + c_{n-d+1} + \cdots + c_{n-1}$ request slots in the previous

d frames and we are still free to choose c_n, which makes that the number of arriving packets in the next d frames is given by

$$\sum_{k=1}^{d}\sum_{i=1}^{c_{n-k}} Y_{n-ki} + \sum_{i=1}^{c_n} Y_{ni}. \tag{8.21}$$

Ideally, there will be $f - c_n$ packets at the beginning of each frame, so that in each frame all waiting packets can be transmitted. In that case, we would have

$$X_n = f - c_n; \qquad \sum_{k=1}^{d}\sum_{i=1}^{c_{n-k}} Y_{n-ki} + \sum_{i=1}^{c_n} Y_{ni} = (d+1)\Lambda. \tag{8.22}$$

In reality, this wil not be the case, but we will take these values as a benchmark. So, we aim at choosing c_n such that

$$X_n - (f - c_n) + \sum_{k=1}^{d}\sum_{i=1}^{c_{n-k}} Y_{n-ki} + \sum_{i=1}^{c_n} Y_{ni} \approx (d+1)\Lambda. \tag{8.23}$$

This benchmark provides a useful scheduling strategy when we replace the Y_{ni} in (8.23) by their expectation μ_Y. Some rewriting then gives the value $\bar{c}_n$ as a target level for c_n, i.e.

$$
\begin{aligned}
\bar{c}_n &= \frac{1}{1+\mu_Y}\left[f + (d+1)\Lambda - X_n - \mu_Y \sum_{k=1}^{d} c_{n-k} \right] \\
&= c^* + \frac{1}{1+\mu_Y}\left[(d+1)\Lambda - X_n - \mu_Y \sum_{k=1}^{d} c_{n-k} \right].
\end{aligned} \tag{8.24}
$$

To make sure that c_n is integer-valued, and that all unused data slots are turned into request slots, we then choose c_n according to

$$c_n = \max(0, \lfloor \bar{c}_n \rfloor, f - X_n). \tag{8.25}$$

8.4   Numerical Assessment

In order to assess the merit of the two scheduling strategies, we have carried out a number of simulations. As before, $f = 18$, and we vary Λ, d, and c. We let Y follow a Poisson distribution. In all simulation results presented below, the performance measures have been evaluated on an interval of 1,000,000 frames, after an initial warm-up period of 200,000 frames.

In our assessment, we investigate whether our scheduling strategies succeed in reducing the autocorrelation. Additionally we compare the average queue lengths under both static and adaptive scheduling.

8.4.1 Impact on the Autocorrelation

The intention to reduce the autocorrelation of the arrival process was the corner stone of the adaptive scheduling strategy. Here, we investigate the extent to which the strategy, implemented via (8.25) achieves this. Moreover, we compare this with the autocorrelation achieved with a static scheduling strategy with a given, fixed, value for c.

In Figs. 8.7, 8.9, and 8.11 we have plotted $\mathbb{E}R^d$ for static and adaptive scheduling, for Poisson arrivals and $d = 1, 3, 25$.

Note that adaptive scheduling very well achieves its aim and drives the autocorrelation to zero, up to a given traffic intensity. The higher the transmission delay becomes, the more (relatively) the correlation term is lowered by adaptive scheduling. For $d = 25$, the correlation term for adaptive scheduling is almost negligible. This can be explained as follows. The adaptive scheduling determines the appropriate number of request slots by estimating the number of packets that will arrive in the future. Denote the total number of request slots scheduled in the d previous frames by m. The estimated number of future arrivals is then $m\mu_Y$. Hence, the larger d, the larger m, and the more precise the estimation of the number of future arrivals will be.

8.4.2 Comparison

We now compare the performance of adaptive scheduling with the performance of static scheduling, see Figs. 8.8, 8.10, and 8.12. The performance of the

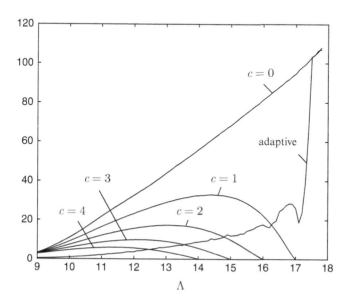

Fig. 8.7. $\mathbb{E}R^d, d = 1$

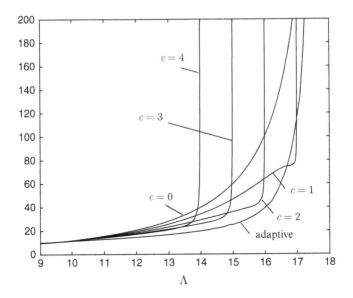

Fig. 8.8. $\mathbb{E}X^d$, $d = 1$

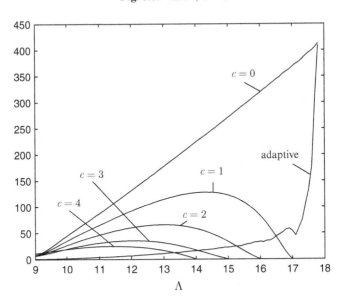

Fig. 8.9. $\mathbb{E}R^d$, $d = 3$

scheduling strategy with fixed c is indicated by various curves, labeled with the corresponding c. Clearly, each of these curves has an asymptote at $18 - c$, as at most $18 - c$ data packets can be transported if c slots are guaranteed to the request process.

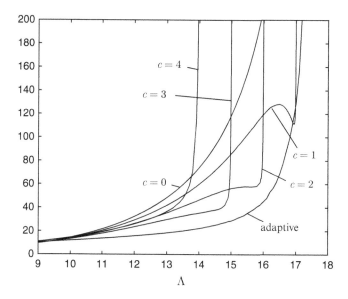

Fig. 8.10. $\mathbb{E}X^d, d = 3$

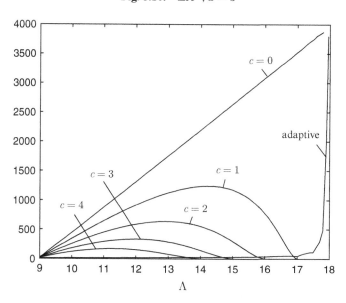

Fig. 8.11. $\mathbb{E}R^d, d = 25$

First note that the best performance of static scheduling is given by the lower envelope of the curves labeled with the c's. As clearly visible, adaptive scheduling performs better. Also, the difference with static scheduling tends to get bigger for increasing values of d, which can be largely attributed to the reduction

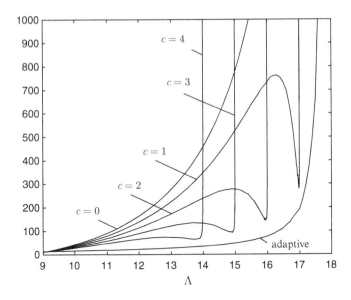

Fig. 8.12. $\mathbb{E}X^d$, $d = 25$

in the autocorrelation as discussed in the previous subsection. So practically in all cases the adaptive scheduling performs well, in the sense that it minimises the mean queue length for almost all values of Λ. An exception is for Λ ranging from 16.7 to 17 for $d = 1$. In this case, regular scheduling with $c = 1$ gives a smaller mean queue length.

For increasing values of d, the relative performance of the adaptive scheduling becomes better. The reason for this can be seen from the figures that display the correlation term. The higher the transmission delay becomes, the more (relatively) the correlation term is lowered by adaptive scheduling.

Figure 8.13 displays the variance of the stationary queue length for Poisson arrivals and $d = 3$. What strikes is that the figure is quite similar to Fig. 8.10, including the good performance of adaptive scheduling and the non-monotonic behaviour. We have seen similar behaviour for geometric arrivals.

8.5 Conclusion

We have introduced the delayed bulk service queue as a model for the size of the data queue in a reservation process. Using a technique by Kingman [95], a bounding technique from Denteneer et al. [50], and a fluid approximation, we have obtained an approximation to the mean data-queue length. Simulations have shown that the approximation is satisfactory in case the data follow a geometric or Poisson distribution, $d > 0$, and $c = 0$. However, the approximation breaks down in heavy traffic conditions for the case that $c > 0$.

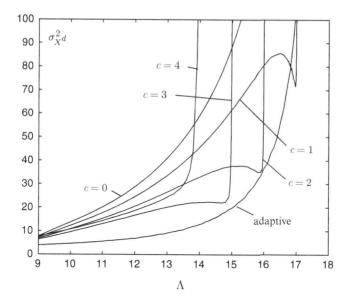

Fig. 8.13. $\sigma^2_{X^d}, d = 3$

Both the approximation and simulations revealed a number of interesting properties of the delayed bulk service queue. Firstly, we showed that the mean data-queue length is increasing in d and grows without limits for d tending to infinity. Secondly, and rather remarkably, we showed that for $d > 0$ and $c > 0$, the mean data-queue length is not monotonic in the traffic intensity. Thus, the mean data-queue length first grows with the mean traffic intensity. However, as the traffic intensity approaches the stability bound, the mean data-queue length decreases substantially.

Thirdly, we consider the heavy-traffic limit. The correlation term will vanish in the heavy-traffic limit where μ_Y approaches $(f - c)/c$. Thus, Theorem 8.1 implies that there exists a heavy-traffic limit in case $c > 0$. In fact, for $c > 0$, the heavy-traffic limit for the delayed flexible boundary model equals the heavy traffic limit for the ordinary bulk service queue and will be dominated by the term $c\sigma^2_Y/(2(f - c - c\mu_Y))$. In case $c = 0$ there is no stability bound, and the expected inflow following empty periods equals $f\mu_Y$ which always increases in μ_Y.

Fourthly, we showed that (8.4) admits a static scheduling strategy. This scheduling strategy relies on the fact that the optimisation of the mean data-queue length with respect to c yields a non-trivial function of the traffic intensity. These findings are in line with simulation results in [78, 150].

Our investigations also led us to propose an alternative, adaptive, scheduling strategy. This adaptive strategy determines the number of request slots in frame

n based on the queue length at the beginning of frame n and the request slots scheduled in the d previous frames. It thus takes the propagation delay into account. Numerical examples showed that the adaptive scheduling performs much better than the static scheduling, due to the fact that adaptive scheduling better succeeds in reducing the correlation of the input process.

APPENDIX 8.A: Proof of Theorem 8.1

Define

$$S_{n-d} = \sum_{i=1}^{c+M_{n-d}} Y_{n-d,i}. \tag{8.A.1}$$

It holds that $X_n - s = P_n - M_n = X_{n+1} - S_{n-d} - M_n$. Take expectations, limits for $n \to \infty$, and rearrange to obtain

$$\mathbb{E}(M^d) = (s - c\mu_Y)/(1 + \mu_Y). \tag{8.A.2}$$

Next, we use that $P_n = X_{n+1} - S_{n-d}$ together with (8.6) to obtain

$$
\begin{aligned}
(X_n - s)^2 &= P_n^2 + M_n^2 \\
&= (X_{n+1} - S_{n-d})^2 + M_n^2 \\
&= X_{n+1}^2 - 2X_{n+1}S_{n-d} + S_{n-d}^2 + M_n^2 \\
&= X_{n+1}^2 - 2(P_n + S_{n-d})S_{n-d} + S_{n-d}^2 + M_n^2 \\
&= X_{n+1}^2 - 2P_n S_{n-d} - S_{n-d}^2 + M_n^2. \tag{8.A.3}
\end{aligned}
$$

It follows in particular that

$$2sX_n = s^2 + X_n^2 - X_{n+1}^2 + 2P_n S_{n-d} + S_{n-d}^2 - M_n^2. \tag{8.A.4}$$

Furthermore, it holds that

$$\mathbb{E}\left(S_{n-d}^2\right) = (c + \mathbb{E}(M_{n-d}))\sigma_Y^2 + \left(c^2 + 2c\mathbb{E}(M_{n-d}) + \mathbb{E}(M_{n-d}^2)\right)\mu_Y^2, \tag{8.A.5}$$

and

$$
\begin{aligned}
\mathbb{E}(P_n S_{n-d}) &= c\mathbb{E}(P_n)\mu_Y + \mathbb{E}(P_n M_{n-d})\mu_Y \\
&= c\mathbb{E}(X_n - s + M_n)\mu_Y + \mathbb{E}(P_n M_{n-d})\mu_Y. \tag{8.A.6}
\end{aligned}
$$

Take expectations in (8.A.4), substitute (8.A.5) and (8.A.5), take limits for $n \to \infty$, and rearrange to obtain

$$
\begin{aligned}
\mathbb{E}(X^d)2(s - c\mu_Y) &= (s - c\mu_Y)^2 + 2c\mathbb{E}(M^d)(\mu_Y + \mu_Y^2) + \mathbb{E}(M^d)\sigma_Y^2 \\
&\quad + c\sigma_Y^2 + \mathbb{E}((M^d)^2)(\mu_Y^2 - 1) + 2\mathbb{E}(R^d)\mu_Y,
\end{aligned}
$$

with $\mathbb{E}(R^d)$ given by (8.8). Finally, substituting (8.A.2) yields (8.7).   □

APPENDIX 8.B: Stationary Distribution

Let $Y(z) = \sum_{k=0}^{\infty} \mathbb{P}(Y = k)z^k$ be the pgf of Y and define $y_k^j = \mathbb{P}(Y_1 + \cdots + Y_j = k)$, where Y_i i.i.d. as Y for all i. Denote by μ_Y and σ_Y^2 the mean and variance of Y. Let $M_n = (s - X_n)^+$, where $x^+ := \max(0, x)$. From (8.4) it is clear that $Z_n := \{(X_n, M_{n-1}, \ldots, M_{n-d})\}$ is a Markov chain. The variables $M_{n-1}, \ldots, M_{n-d}$ constitute our memory, in the sense that these variables keep track of all reservation efforts of which the outcome is still not known. We henceforth assume that this Markov chain is irreducible and aperiodic; for example, this holds when $\mathbb{P}(Y = k) > 0$ for all $k \geq 0$. The ergodicity condition is formulated in the following lemma.

LEMMA 8-8.B.1 *The Markov chain Z_n is ergodic if*

$$c\mu_Y - s < 0. \tag{8.B.1}$$

Proof By partitioning the state space of the Markov chain Z_n into levels i, where level i is the subset of states for which the queue length is i, $i = 0, 1, \ldots$, it is readily seen that the probability transition matrix is an M/G/1-type stochastic matrix, see [133]. Hence, the ergodicity condition is the usual condition stating that the average drift should be negative, which in this case reduces to inequality (8.B.1). For an alternative proof of (8.B.1), based on Foster's criterion, the reader is referred to [52]. □

In the sequel we assume that (8.B.1) is satisfied. The Markov chain Z_n then has a unique stationary distribution

$$\pi(k, m_1, \ldots, m_d) = \lim_{n \to \infty} \mathbb{P}(X_n = k, M_{n-1} = m_1, \ldots, M_{n-d} = m_d), \tag{8.B.2}$$

where $k \geq 0$ and $m_i \in \{0, \ldots, s\}$ for all i. Let X denote a random variable distributed according to the stationary queue length distribution

$$\pi(k) := \mathbb{P}(X = k) = \lim_{n \to \infty} \mathbb{P}(X_n = k). \tag{8.B.3}$$

The probability generating function of X is then given by

$$G(z) = \sum_{m_1=0}^{s} \cdots \sum_{m_d=0}^{s} G_{m_1,\ldots,m_d}(z), \tag{8.B.4}$$

where

$$G_{m_1,\ldots,m_d}(z) = \sum_{k=0}^{\infty} \pi(k, m_1, \ldots, m_d)z^k. \tag{8.B.5}$$

In the analysis below we make use of the normalisation condition

$$\sum_{k=0}^{s-1} \sum_{m_1=0}^{s} \cdots \sum_{m_d=0}^{s} \pi(k, m_1, \ldots, m_d)(s - k) = \frac{s - c\mu_Y}{1 + \mu_Y}, \tag{8.B.6}$$

which can be explained as follows. The left-hand side of (8.B.6) clearly expresses the additional reservation effort per frame. On average, the guaranteed reservation effort brings per frame $c\mu_Y$ of new packets to the queue, each packet requiring one slot. Hence, $c\mu_Y$ of the remaining s slots per frame are spent on serving these packets. This leaves $s - c\mu_Y$ slot, of which only $(s - c\mu_Y)/(1 + \mu_Y)$ slots can be spent on additional reservation (since one reservation slot results on average in μ_Y new packets).

Case $d = 0$

We start from the balance equations (for $k = 0, 1, \ldots$)

$$\pi(k) = \sum_{k'=s}^{k+s} \pi(k')y_{k-k'+s}^c + \sum_{k'=0}^{s-1} \pi(k')y_k^{c+s-k'}. \tag{8.B.7}$$

Multiplying both sides of (8.B.7) by z^k, summing over all values of k, and rearranging terms yields

$$G(z) = \frac{Y(z)^c \sum_{k=0}^{s-1} \pi(k)(z^s Y(z)^{s-k} - z^k)}{z^s - Y(z)^c}. \tag{8.B.8}$$

The expression (8.B.8) is of indeterminate form, but the s boundary probabilities $\pi(0), \ldots, \pi(s - 1)$ can be determined by consideration of the zeros of the denominator in (8.B.8) that lie on or within the unit circle. The following lemma is a restatement of Theorem 6.1.

LEMMA 8-8.B.2 *If $c\mu_Y < s$ and $\mathbb{P}(Y = 0) > 0$, it holds that $z^s = Y(z)^c$ has s roots on or within the unit circle.*

Denote the s roots of $z^s = Y(z)^c$ in $|z| \leq 1$ by $z_0 = 1, z_1, \ldots, z_{s-1}$. Since the function $G(z)$ is finite on and inside the unit circle, the numerator of the right-hand side of (8.B.8) needs to be zero for each of the s roots, i.e., the numerator should vanish at the exact points where the denominator of the right-hand side of (8.B.8) vanishes. For $z_0 = 1$, this is trivial, so Lemma 8-8.B.2 and (8.B.8) lead to $s - 1$ (non-trivial) equations in terms of the s boundary probabilities. The final equation follows from the normalisation condition (8.B.6).

Case $d = 1$

In this case we include one memory variable M_{n-1} into our state description, and we distinguish between the balance equations for states (k, m_1) for which $m_1 = 0$

$$\pi(k, 0) = \sum_{i=0}^{s} \sum_{k'=s}^{k+s} \pi(k', i)y_{k-k'+s}^{c+i}, \tag{8.B.9}$$

and for which $m_1 \in \{1, \ldots, s\}$

$$\pi(k, m_1) = \sum_{i=0}^{s} \pi(s - m_1, i)y_k^{c+i}. \tag{8.B.10}$$

Multiplying both sides of (8.B.9) and (8.B.10) by z^k and summing over all values of k yields

$$G_0(z) = z^{-s} \sum_{i=0}^{s} \left(G_i(z) - \sum_{k=0}^{s-1} \pi(k, i)z^k \right) Y(z)^{c+i}, \tag{8.B.11}$$

and

$$G_{m_1}(z) = \sum_{i=0}^{s} \pi(s - m_1, i)Y(z)^{c+i}, \quad m_1 \in \{1, \ldots, s\}, \tag{8.B.12}$$

respectively. Upon substituting (8.B.12) into (8.B.11) and rearranging terms we find

$$G_0(z) = \frac{Y(z)^c \sum_{i=0}^{s} \sum_{k=0}^{s-1} \pi(k, i)(Y(z)^{s+c+i-k} - z^k Y(z)^i)}{z^s - Y(z)^c}. \tag{8.B.13}$$

Hence, (8.B.12) and (8.B.13) still contain $s(s+1)$ boundary probabilities

$$\pi(k, m_1), \quad k = 0, \ldots, s-1, \quad m_1 = 0, \ldots, s,$$

which should be determined. We therefore match these unknowns by equally many equations: s^2 equations follow from (8.B.10), $s - 1$ equations follow from Lemma 8-8.B.2 and (8.B.13), and a final equation is provided by the normalisation condition (8.B.6).

Case $d = 2$

We now include two memory variables M_{n-1} and M_{n-2} into our state description, but for the balance equations we only need to distinguish between the states (k, m_1, m_2) with $m_1 = 0$ and $m_1 \in \{1, \ldots, s\}$. We get

$$\pi(k, 0, m_2) = \sum_{i=0}^{s} \sum_{k'=s}^{k+s} \pi(k', m_2, i)y_{k-k'+s}^{c+i}, \tag{8.B.14}$$

and for $m_1 \in \{1, \ldots, s\}$

$$\pi(k, m_1, m_2) = \sum_{i=0}^{s} \pi(s - m_1, m_2, i)y_k^{c+i}, \tag{8.B.15}$$

and so we obtain for $m_2 \in \{0, \ldots, s\}$

$$G_{0,m_2}(z) = z^{-s} \sum_{i=0}^{s} \left(G_{m_2,i}(z) - \sum_{k=0}^{s-1} \pi(k, m_2, i) z^k \right) Y(z)^{c+i}, \quad \text{(8.B.16)}$$

and for $m_1 \in \{1, \ldots, s\}$ and $m_2 \in \{0, \ldots, s\}$

$$G_{m_1,m_2}(z) = \sum_{i=0}^{s} \pi(s - m_1, m_2, i) Y(z)^{c+i}. \quad \text{(8.B.17)}$$

Upon rearranging terms in (8.B.16) for $m_2 = 0$ we get

$$G_{0,0}(z) = \frac{Y(z)^c \sum_{i=1}^{s}(G_{0,i}(z)Y(z)^i - \sum_{i=0}^{s}\sum_{k=0}^{s-1}\pi(k,0,i)z^k Y(z)^i)}{z^s - Y(z)^c}.$$

$$\text{(8.B.18)}$$

Equations (8.B.16)–(8.B.18) contain $s(s + 1)^2$ boundary probabilities

$$\pi(k, m_1, m_2), \qquad k \in \{0, \ldots, s-1\}, \quad m_1, m_2 \in \{0, \ldots, s\}, \quad \text{(8.B.19)}$$

which can again be matched by equally many equations. In this case, $s^2(s+1)$ equations follow from (8.B.15) for $k \in \{0, \ldots, s-1\}$, $m_1 \in \{1, \ldots, s\}$, $m_2 \in \{0, \ldots, s\}$, $s-1$ equations follow from Lemma 8-8.B.2 and (8.B.18) (using (8.B.16) and (8.B.17)), and one equation follows from the normalisation condition (8.B.6). So we need an extra s^2 equations. For this, we consider the probabilities $\pi(k, 0, m_2)$, $k \in \{0, \ldots, s-1\}$, $m_2 \in \{1, \ldots, s\}$, see Fig. 8.B.1. Note that these probabilities can be expressed through (8.B.14) in terms of the probabilities $\pi(k', m_2, i)$, $k' \in \{s, \ldots, 2s-1\}$, $m_2 \in \{1, \ldots, s\}$, $i \in \{0, \ldots, s\}$, only. Each of the latter probabilities can be written in terms of the boundary probabilities through (8.B.15), yielding s^2 equations.

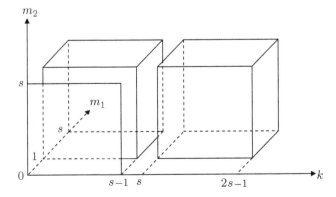

Fig. 8.B.1. The states corresponding to the boundary probabilities (8.B.19) and the additional probabilities $\pi(k', m_2, i)$, $k' \in \{s, \ldots, 2s-1\}$, $m_2 \in \{1, \ldots, s\}$, $i \in \{0, \ldots, s\}$

General Case

It might be clear from the analysis for $d = 2$ that we can take a similar approach for $d = 3, 4, \ldots$. We start from the balance equations

$$\pi(k, 0, m_2, \ldots, m_d) = \sum_{i=0}^{s} \sum_{k'=s}^{k+s} \pi(k', m_2, \ldots, m_d, i) y_{k-k'+s}^{c+i}, \qquad (8.B.20)$$

and for $m_1 \in \{1, \ldots, s\}$

$$\pi(k, m_1, m_2, \ldots, m_d) = \sum_{i=0}^{s} \pi(s - m_1, m_2, \ldots, m_d, i) y_k^{c+i}, \qquad (8.B.21)$$

and obtain

$$\frac{G_{0,m_2,\ldots,m_d}(z)}{Y(z)^{c+i}} = z^{-s} \sum_{i=0}^{s} \left(G_{m_2,\ldots,m_d,i}(z) - \sum_{k=0}^{s-1} \pi(k, m_2, \ldots, m_d, i) z^k \right),$$

$$(8.B.22)$$

and for $m_1 \in \{1, \ldots, s\}$

$$G_{m_1,\ldots,m_d}(z) = \sum_{i=0}^{s} \pi(s - m_1, m_2, \ldots, m_d, i) Y(z)^{c+i}. \qquad (8.B.23)$$

For $m_2 = m_3 = \ldots = m_d = 0$ we get from (8.B.22)

$$G_{0,\ldots,0}(z) = \frac{\sum_{i=1}^{s} \left(G_{0,\ldots,0,i}(z) - \sum_{k=0}^{s-1} \pi(k, 0, \ldots, 0, i) z^k \right) Y(z)^{c+i}}{z^s - Y(z)^c}.$$

$$(8.B.24)$$

We should then still determine the $s(s+1)^d$ boundary probabilities

$$\pi(k, m_1, \ldots, m_d), \qquad k \in \{0, \ldots, s-1\}, \qquad m_1, \ldots, m_d \in \{0, \ldots, s\}$$
$$(8.B.25)$$

and so we need equally many equations. Equation (8.B.21) immediately yields $s^2(s+1)^{d-1}$ equations, and we thus search for an extra $s(s+1)^{d-1}$ equations.

Consider (8.B.20) for $m_2 \in \{1, \ldots, s\}$. The probabilities on the right-hand side of (8.B.20) can be written in terms of the boundary probabilities through (8.B.21), which yields $s^2(s+1)^{d-2}$ equations. Likewise, (8.B.20) for $m_2 = 0$ and $m_3 \in \{1, \ldots, s\}$ yields $s^2(s+1)^{d-3}$ equations, and we can repeat this trick all the way up to (8.B.20) for $m_2 = \cdots = m_{d-1} = 0$, $m_d \in \{1, \ldots, s\}$ (which yields s^2 equations). Altogether, this leads to

$$s^2(s+1)^{d-2} + s^2(s+1)^{d-3} + \cdots + s^2$$

equations. Together with the $s-1$ equations from Lemma 8-8.B.2 and (8.B.24), and the normalisation condition (8.B.6), we then have exactly $s(s+1)^{d-1}$ equations.

PART IV

SHARED SERVICE CAPACITY

Chapter 9

A TANDEM QUEUE
WITH COUPLED PROCESSORS

In this chapter we investigate the two-stage tandem queue with coupled processors, which has been suggested as a model for a reservation mechanism in Sect. 2.4. It is assumed that jobs arrive at the first station according to a Poisson process and require service at both stations before leaving the system. The amounts of work that a job requires at each of the stations are independent, exponentially distributed random variables. When both stations are nonempty, the total service capacity is shared between the stations according to fixed proportions. When one of the stations becomes empty, the total service capacity is given to the nonempty station.

We study the two-dimensional Markov process that represents the numbers of jobs at the two stations. The problem of finding the generating function of the stationary distribution can be reduced to two different Riemann–Hilbert boundary value problems. Although both problems yield a complete analytical solution, they have different features from the numerical viewpoint. We discuss the similarities and differences between the two problems, and relate them to the computational aspects of obtaining performance measures.

9.1 Introduction

One application of the tandem queue with coupled processors is a cable access network regulated by a reservation mechanism, and we refer to Sect. 2.4 for an elaboration of this point of view. Another application of the model would be an assembly line for which two operations on each job must be performed using a limited service capacity. By coupling the service at each of the operations, and thus using the service capacity of an operation for which no jobs are waiting for the other operation, imbalance in the assembly line can be reduced and the throughput can be increased, see, e.g. Andradottir et al. [6].

Resing and Örmeci [148] have shown for the two-dimensional Markov process representing the numbers of jobs at the two stations, that the problem of finding the bivariate generating function of the stationary distribution can be reduced to a Riemann–Hilbert boundary value problem. In [148] the issue of how to obtain performance measures has not been discussed. In general, obtaining performance measures from the formal solution of a Riemann–Hilbert boundary value problem is not straightforward. In this chapter we discuss the numerical issues that arise when computing performance measures. In particular, we consider the fraction of time a station is empty and the mean stationary queue length at a station. The reduction of the problem of finding the generating function to a boundary value problem usually follows from considering a specific zero-set of the kernel of the functional equation. This can be done in more than one way. We discuss, next to the zero-set considered in [148], one other zero-set that leads to a second Riemann–Hilbert boundary value problem. From the analytical viewpoint, the second formulation has little added value, since solving either one of the two problems gives a full solution to the model. However, in determining performance measures numerically, the two problems have different features.

We describe the model and the key functional equation for the model in Sect. 9.2. In Sect. 9.3 we analyse the kernel of the functional equation. The results are used in Sects. 9.4 and 9.5 to reduce the problem of solving the functional equation to two different Riemann–Hilbert boundary value problems. Specific attention is paid to determining the conformal map that is required for the solution of the second Riemann–Hilbert boundary value problem. In Sect. 9.6 we derive some performance measures for the model and we discuss issues that arise when numerically determining the performance measures from the formal solutions of the Riemann–Hilbert boundary value problems. Among other things, we show that we can determine the performance measures for the whole set of allowed parameter values. Finally, we give some numerical results in Sect. 9.6.5.

9.2   Model Description

Consider a two-stage tandem queue, where jobs arrive at queue 1 according to a Poisson process with rate λ, each job demanding service at both queues before leaving the system. Each job requires an exponential amount of work with parameter ν_j at station j, $j = 1, 2$. The total service capacity of the two stations together is fixed. Without loss of generality we assume that this total service capacity equals one unit of work per time unit. Whenever both stations are nonempty, a proportion p of the capacity is allocated to station 1, and the remaining part $1-p$ is allocated to station 2. Thus, when there is at least one job at each station, the departure rate of jobs at station 1 is $\nu_1 p$ and the departure rate of jobs at station 2 is $\nu_2(1-p)$. Here we assume that $0 < p < 1$, so that there is a real capacity sharing between the two stations. For the cases $p = 0$ and $p = 1$, the system can be seen as a tandem queue with a single server moving between the two queues and giving priority to one of the queues. The solutions for these cases are given in Resing and Örmeci [148].

When one of the stations becomes empty, the total service capacity is allocated to the nonempty station. Hence, the departure rate at that station, station j say, is temporarily increased to ν_j. With $X_j(t)$ the number of jobs at station j at time t, the two-dimensional process $\{(X_1(t), X_2(t)), t \geq 0\}$ is a Markov process. The condition under which this Markov process has a unique stationary distribution is given by

$$\frac{\lambda}{\nu_1} + \frac{\lambda}{\nu_2} < 1. \tag{9.1}$$

This can be explained by the fact that, independent of p, the two stations together always work at capacity 1 (if there is work in the system), and that $\lambda/\nu_1 + \lambda/\nu_2$ equals the amount of work brought into the system per time unit. We henceforth assume that the ergodicity condition is satisfied.

Let us denote by $\pi(n, k)$ the stationary probability of having n customers at station 1 and k customers at station 2, i.e. $\pi(n, k) = \lim_{t\to\infty} \mathbb{P}(X_1(t) = n, X_2(t) = k)$. The following set of balance equations can then be derived:

$$\lambda\pi(0, 0) = \nu_2\pi(0, 1),$$
$$(\lambda + \nu_2)\pi(0, 1) = \nu_1\pi(1, 0) + \nu_2\pi(0, 2),$$
$$(\lambda + \nu_2)\pi(0, k) = p\nu_1\pi(1, k-1) + \nu_2\pi(0, k+1), \quad k \geq 2,$$

and for $n \geq 1$

$$(\lambda + \nu_1)\pi(n, 0) = \lambda\pi(n-1, 0) + (1-p)\nu_2\pi(n, 1),$$
$$\lambda + p\nu_1 + (1-p)\nu_2\pi(n, 1) = \lambda\pi(n-1, 1) + \nu_1\pi(n+1, 0)$$
$$+ (1-p)\nu_2\pi(n, 2),$$
$$\lambda + p\nu_1 + (1-p)\nu_2\pi(n, k) = \lambda\pi(n-1, k) + p\nu_1\pi(n+1, k-1)$$
$$+ (1-p)\nu_2\pi(n, k+1), \quad k \geq 2.$$

We define the joint probability generating function

$$P(x, y) := \sum_{n\geq 0}\sum_{k\geq 0} \pi(n, k)x^n y^k, \quad |x| \leq 1, |y| \leq 1,$$

which is, for every fixed y, regular for $|x| < 1$ and continuous for $|x| \leq 1$. A similar statement holds for x and y interchanged. From the balance equations it follows that $P(x, y)$ satisfies the functional equation

$$h_1(x, y)P(x, y) = h_2(x, y)P(x, 0) + h_3(x, y)P(0, y) + h_4(x, y)P(0, 0), \tag{9.2}$$

where

$$h_1(x, y) = (\lambda + p\nu_1 + (1 - p)\nu_2)xy - \lambda x^2 y - p\nu_1 y^2 - (1 - p)\nu_2 x,$$
$$h_2(x, y) = (1 - p)[\nu_1 y(y - x) + \nu_2 x(y - 1)],$$
$$h_3(x, y) = p[\nu_2 x(1 - y) + \nu_1 y(x - y)],$$
$$h_4(x, y) = p\nu_2 x(y - 1) + (1 - p)\nu_1 y(x - y).$$

The constant $P(0, 0)$ can be determined by substituting

$$x = \gamma(y) := \nu_1 y^2 / (\nu_1 y - \nu_2 y + \nu_2)$$

into (9.1). For this choice of x, both $h_2(x, y)$ and $h_3(x, y)$ equal zero, and hence (9.1) reduces to

$$P(\gamma(y), y) = \frac{h_4(\gamma(y), y)}{h_1(\gamma(y), y)} P(0, 0). \tag{9.3}$$

Letting $y \uparrow 1$ in (9.3), we obtain $P(0, 0) = 1 - \lambda/\nu_1 - \lambda/\nu_2$. This result can again be explained by the fact that, independent of p, the two stations together always work at capacity 1 (if there is work in the system), and that $\lambda/\nu_1 + \lambda/\nu_2$ equals the amount of work brought into the system per time unit.

9.3   Analysis of the Kernel

In the analysis of the functional equation (9.1) a crucial role is played by the kernel $h_1(x, y)$. Due to the regularity properties of $P(x, y)$, for each pair (x, y) on or within the unit circle for which $h_1(x, y)$ equals zero, the right-hand side of (9.1) must vanish. This provides us with a relation between the unknown functions $P(0, y)$ and $P(x, 0)$. From the observation that $h_1(x, y)$ is a polynomial in either x or y, we can construct two Riemann–Hilbert boundary value problems, one for the function $P(x, 0)$ and one for the function $P(0, y)$.

Blanc [18] has investigated the transient behaviour of the ordinary tandem queue without coupled processors, for which the kernel $h_1(x, y)$ is of exactly the same form. Since Blanc has studied $h_1(x, y)$ as a polynomial in y, most of the results presented in this section stem from his work. Using these results, the problem of finding the stationary queue length distribution can be reduced to a Riemann–Hilbert boundary value problem for $P(0, y)$, as presented in Sect. 9.4. In Sect. 9.5 we derive a Riemann–Hilbert boundary value problem for $P(x, 0)$.

We introduce

$$r_1 = \frac{\lambda}{p\nu_1}, \qquad r_2 = \frac{\lambda}{(1 - p)\nu_2},$$

as the loads on each of the stations if they would work in isolation (no coupling). For notational convenience, we also introduce

$$\hat{r} = 1 + \frac{1}{r_1} + \frac{1}{r_2},$$

such that

$$h_1(x, y) = \lambda \left[\hat{r}xy - x^2 y - \frac{1}{r_1} y^2 - \frac{1}{r_2} x \right]. \tag{9.4}$$

Observe that $h_1(x, y)$ is, for each x, a polynomial of degree 2 in y. We thus have that for every value of x there are two possible values of y, say $y_1(x)$ and $y_2(x)$, such that $h_1(x, y_1(x)) = h_1(x, y_2(x)) = 0$. These can be described by the two-valued function

$$y(x) = \frac{r_1}{2} [s_1(x) \pm \sqrt{D_1(x)}], \tag{9.5}$$

where

$$s_1(x) = (\hat{r} - x)x, \quad D_1(x) = s_1(x)^2 - \frac{4x}{r_1 r_2}.$$

We then obtain the following result:

LEMMA 9.1 *The algebraic function* $y(x)$, *defined by* $h_1(x, y(x)) = 0$, *has four real branch points* $0 = x_1 < x_2 \le 1 < x_3 < x_4$.

Proof The branch points of $y(x)$ are zeros of the discriminant $D_1(x)$. Clearly, $D_1(0) = 0$, $\lim_{x \downarrow 0} D_1(x) < 0$, $D_1(1) \ge 0$, $D_1(\hat{r}) < 0$ and $\lim_{x \to \infty} D_1(x) = \infty$. Furthermore, if $D_1(1) = 0$ (i.e. $r_1 = r_2 < 1$) then $D_1'(1) > 0$. $\qquad \square$

For later use, we present the following lemma which shows that the mapping $y(x)$ for $x \in [0, x_2]$ gives rise to a smooth and closed contour L, see Fig. 9.1.

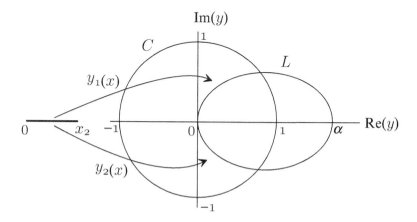

Fig. 9.1. The mapping $y = y(x) : [0, x_2] \to L$

LEMMA 9.2 *For each $x \in [0, x_2]$, $y(x)$ lies on the closed contour L, which is symmetric with respect to the real line, and defined by*

$$|y|^2 = \frac{r_1}{2r_2} \left(\hat{r} - \sqrt{\hat{r}^2 - 8\mathrm{Re}(y)/r_1} \right). \tag{9.6}$$

It further holds that

$$|y|^2 \le \frac{r_1}{r_2} x_2. \tag{9.7}$$

Proof For $x \in [0, x_2]$, $D_1(x)$ is negative, so $y_1(x)$ and $y_2(x)$ are complex conjugates. It also follows that

$$\mathrm{Re}(y) = \frac{r_1}{2}(\hat{r} - x)x. \tag{9.8}$$

Furthermore, from $h_1(x, y(x)) = 0$ we have $|y|^2 = r_1 x/r_2 \le r_1 x_2/r_2$. Since (9.8) is a quadratic equation in x, substituting one of the two solutions into $|y|^2 = r_1 x/r_2$ yields (9.6). Of course, we choose the solution of x for which $y(0) = 0$ and $y(x_2) = \sqrt{r_1 x_2/r_2}$ lie on the contour. $\square$

We will henceforth denote the interior of L by L^+, and set

$$\alpha := y(x_2) = \sqrt{r_1 x_2/r_2}, \tag{9.9}$$

representing the point on L with the largest modulus. With respect to α, the following assertions hold.

LEMMA 9.3 *If $r_1 = r_2$, then $\alpha = 1$. If $r_1 < r_2$, then $\alpha < 1$. If $r_1 > r_2$, then $\alpha > 1$.*

Proof For $r_1 = r_2$, we have that $D_1(1) = 0$, so $x_2 = \alpha = 1$. For $r_1 < r_2$, knowing $x_2 < 1$, it follows that $\alpha < 1$. For $r_1 > r_2$, knowing $x_2 < 1$, we have that $D_1(r_2/r_1) < 0$ since $r_2 + r_2(1 - r_2)/r_1 < 1$, and thus $r_2/r_1 < x_2$ and $\alpha > 1$. $\square$

We note that $\alpha = 1$ (respectively $\alpha < 1$, $\alpha > 1$) implies $1 \in L$ (respectively $1 \notin L \cup L^+$, $1 \in L^+$), which plays a crucial role in the numerical work to be presented in Sect. 9.6.

Paralleling the approach above, the kernel $h_1(x, y)$ is, for each y, a polynomial of degree 2 in x. Thus for each y there are two possible values of x, say $x_1(y)$ and $x_2(y)$, such that $h_1(x_1(y), y) = h_1(x_2(y), y) = 0$. These can be described by the two-valued function

$$x(y) = \frac{1}{2y} \left[s_2(y) \pm \sqrt{D_2(y)} \right], \tag{9.10}$$

where

$$s_2(y) = \hat{r}y - \frac{1}{r_2}, \quad D_2(y) = s_2(y)^2 - \frac{4y^3}{r_1}.$$

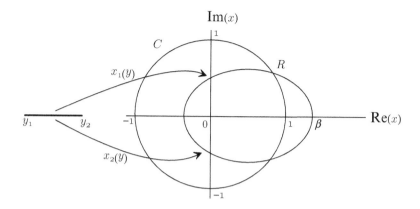

Fig. 9.2. The mapping $x = x(y) : [y_1, y_2] \to R$

The following then holds:

LEMMA 9.4 *The algebraic function $x(y)$ defined by $h_1(x(y), y) = 0$ has three real branch points $0 < y_1 < y_2 \leq 1 < y_3$.*

Proof The branch points of $x(y)$ are zeros of the discriminant $D_2(y)$. Clearly, $D_2(0) = 1/r_2^2 > 0$, $D_2(1) = (1 - 1/r_1)^2 \geq 0$ and $\lim_{y \to \infty} D_2(y) = -\infty$. For $\hat{y} = 1/(r_2\hat{r}) \in (0, 1)$, it holds that $D_2(\hat{y}) = -4\hat{y}^3/r_1 < 0$. Also, if $D_2(1) = 0$ (which implies $r_1 = 1$ and, due to the ergodicity condition, $r_2 < 1$) then $D_2'(1) = 4(1/r_2 - 1) > 0$. $\qquad\square$

We now study the mapping $x(y)$ for $y \in [y_1, y_2]$ in some more detail. This mapping can be shown to give rise to a smooth and closed contour R, as specified in the next lemma and illustrated in Fig. 9.2.

LEMMA 9.5 *For each $y \in [y_1, y_2]$, $x(y)$ lies on the closed and smooth contour R, which is symmetric with respect to the real line, and defined by:*

$$|x|^2 = \frac{1}{r_1 r_2(\hat{r} - 2\mathrm{Re}(x))}, \tag{9.11}$$

$$|x|^2 \leq \frac{y_2}{r_1}. \tag{9.12}$$

Proof Similar to the proof of Lemma 9.2. $\qquad\square$

We set

$$\beta := x(y_2) = \sqrt{y_2/r_1}, \tag{9.13}$$

the point on R with the largest modulus, for which it holds that

LEMMA 9.6 (i) *When either $r_1 = 1$ or $r_2 = 1$ we have that $\beta = 1$. (ii) When both $r_1 < 1$ and $r_2 < 1$ we have that $\beta > 1$. (iii) When either $r_1 > 1$ or $r_2 > 1$ we have that $\beta < 1$.*

Proof (1) If $r_1 = 1$, then $y_2 = 1$ and thus $\beta = 1$. If $r_2 = 1$, then $y_2 = r_1$ and thus $\beta = 1$.

(2) For $\beta > 1$ we should prove that $r_1 < y_2$. Consider the function $f(r_1) := D_2(r_1) = -4r_1^2 + (1 + r_1 + r_1/r_2 - 1/r_2)^2$. The solutions to $f(r_1) = 0$ are given by $r_1 = 1$ and $r_1 = \hat{r}_1 = (1 - r_2)/(1 + 3r_2)$. For $r_1 = \hat{r}_1$ it holds that $r_1 = y_1 < y_2$. Assume that there exists a value $r_1 \in (0, 1)$ for which it holds that $r_1 > y_2$. Then, since y_2 is a continuous function of r_1, there should be a value in $(0, 1)$ other than $\hat{r}_1$ for which $r_1 = y_2$ and hence $f(r_1) = 0$. This is not the case, and thus $r_1 < y_2$ for all values $r_1 \in (0, 1)$.

(3) If $r_1 > 1$, then obviously $r_1 > y_2$ and thus $\beta < 1$. Now assume $r_2 > 1$. Then, for $\beta < 1$ we should prove that $r_1 > y_2$. Note that $f(r_1)$ is positive for all values $r_1 \in (0, 1)$. This implies that $r_1 < y_1$ or $r_1 > y_2$, see the proof of Lemma 9.4. Furthermore, $y_1 < \hat{y} < 1/2$ for $r_2 > 1$ and $r_1 \in (0, 1)$, when $\hat{y}$ as defined in the proof of Lemma 9.4. Hence, for $r_1 \geq 1/2$ it clearly holds that $r_1 > y_1$. Assume that there exists a value $r_1 \in (0, 1)$ for which it holds that $r_1 < y_1$. Then, since y_1 is a continuous function of r_1, there should be a value in $(0, 1)$ for which $r_1 = y_1$ and hence $f(r_1) = 0$. This is not the case, and thus $r_1 > y_2$ for all values $r_1 \in (0, 1)$.   □

We again note that $\beta = 1$ (respectively $\beta < 1$, $\beta > 1$) implies $1 \in R$ (respectively $1 \notin R \cup R^+, 1 \in R^+$), which plays a crucial role in the numerical work to be presented in Sect. 9.6 .

9.4   Boundary Value Problem I

In the previous section we considered the kernel as a polynomial in either y or x, which may lead to the curves L and R, respectively. In this section we describe how the curve L leads to a Riemann–Hilbert boundary value problem for the function $P(0, y)$.

LEMMA 9.7 *The function $P(0, y)$ is regular in the domain L^+ and satisfies for $y \in L$ the condition*

$$\text{Im}(P(0, y)) = \text{Im}\left(-P(0,0)\frac{h_4(r_2|y|^2/r_1, y)}{h_3(r_2|y|^2/r_1, y)}\right). \tag{9.14}$$

Proof For zero-pairs (x, y) of the kernel $h_1(x, y)$ for which $P(x, y)$ is finite, we have

$$h_2(x, y)P(x, 0) + h_3(x, y)P(0, y) + h_4(x, y)P(0, 0) = 0, \tag{9.15}$$

from which it follows that, for those zero-pairs,

$$P(0, y) = \frac{1-p}{p} P(x, 0) - \frac{h_4(x, y)}{h_3(x, y)} P(0, 0). \qquad (9.16)$$

Thus, (9.14) follows from the fact that $P(x, 0)$ is real for $x \in [0, x_2]$ and $|y|^2 = r_1 x / r_2$ for $y \in L$. If $\alpha \leq 1$, L lies entirely within the unit circle. Hence, $P(0, y)$ is regular in L^+. If $\alpha > 1$, $P(0, y(x))$ can be continued analytically over the interval $[0, x_2]$ via (9.15), because $P(x, 0)$ is regular on this interval. Hence, the analytic continuation of $P(0, y)$ is finite at $y = y(x_2)$. Because $P(0, y)$ has a power series expansion at $y = 0$ with positive coefficients, this implies that $P(0, y)$ is regular for $|y| < y(x_2)$ and hence in L^+. $\qquad \square$

Lemma 9.7 shows that the determination of $P(0, y)$ reduces to the determination of the solution of the following Riemann–Hilbert boundary value problem on the contour L: Determine a function $P(0, y)$ such that

1. $P(0, y)$ is regular for $y \in L^+$ and continuous for $y \in L \cup L^+$.

2. $\mathrm{Re}\, (iP(0, y)) = c(y)$, for $y \in L$,

where

$$c(y) = \mathrm{Im}\, \left(P(0, 0) \frac{h_4(r_2|y|^2/r_1, y)}{h_3(r_2|y|^2/r_1, y)} \right).$$

The standard way to solve this type of boundary value problem is to transform the boundary condition (9.14) to a condition on the unit circle, see, e.g. Muskhelishvili [129], p. 108. Denote the unit circle by C and its interior by C^+. We introduce the conformal mapping

$$z = f(y) : L^+ \to C^+, \qquad (9.17)$$

and its inverse

$$y = f_0(z) : C^+ \to L^+. \qquad (9.18)$$

Using these mappings, we can reduce the Riemann–Hilbert problem on L to the following problem: Determine a function $G(z)$ such that

1. $G(z)$ is regular for $z \in C^+$ and continuous for $z \in C \cup C^+$.

2. $\mathrm{Re}\, (iG(z)) = \tilde{c}(z)$, for $z \in C$, where $\tilde{c}(z) = c(f_0(z))$,

which is known as the Dirichlet problem on a circle. Its solution is given by, see Muskhelishvili [129] p. 108,

$$G(z) = -\frac{1}{2\pi} \oint_C \tilde{c}(w) \frac{w+z}{w-z} \frac{\mathrm{d}w}{w} + K_1, \quad z \in C \cup C^+, \qquad (9.19)$$

where K_1 is some constant. In this way, $P(0, y) = G(f(y))$ has been formally determined as

$$P(0, y) = -\frac{1}{2\pi} \oint_C c(f_0(w)) \frac{w + f(y)}{w - f(y)} \frac{dw}{w} + K_1, \quad y \in L \cup L^+. \quad (9.20)$$

We can rewrite the contour integral (9.20) as a real integral on $[0, x_2]$. That is, for $y \in L^+ \cup L$, we have that

$$P(0, y) = -\frac{1}{2\pi} \oint_L c(s) \frac{f(s) + f(y)}{f(s) - f(y)} \frac{df(s)}{f(s)} + K_1 \quad (9.21)$$

$$= \frac{1}{2\pi} \left[\int_0^{x_2} c(y_1(x)) \frac{f(y_1(x)) + f(y)}{f(y_1(x)) - f(y)} \frac{f'(y_1(x))y_1'(x)}{f(y_1(x))} dx \right.$$

$$\left. - \int_0^{x_2} c(y_2(x)) \frac{f(y_2(x)) + f(y)}{f(y_2(x)) - f(y)} \frac{f'(y_2(x))y_2'(x)}{f(y_2(x))} dx \right] + K_1. \quad (9.22)$$

REMARK 9.1 For this specific problem, an explicit expression for the conformal mapping $f(y)$ can be found, see Blanc [18]. It is given by

$$f(y) = \frac{yk(\eta) - \eta k(y)}{yk(\eta) + \eta k(y)}, \quad (9.23)$$

where

$$k(y) = (\alpha - y)\sqrt{r_1 - r_2^2\alpha^2 y},$$

and η is some unspecified constant in the interval $(0, \alpha)$. For our computations we set $\eta = \alpha/2$. With the explicit expression for $f(y)$ we have all ingredients for calculating the integral (9.21), as will be further discussed in Sect. 9.6.

9.5   Boundary Value Problem II

In this section we will show how the second zero-set discussed in Sect. 9.3 that leads to the curve R gives rise to a Riemann–Hilbert problem for the function $P(x, 0)$. The approach is similar to the one followed in Sect. 9.4.

LEMMA 9.8 *The function $P(x, 0)$ is regular in the domain R^+ and satisfies for $x \in R$ the condition*

$$\text{Im}(P(x, 0)) = \text{Im}\left(-P(0, 0) \frac{h_4(x, r_1|x|^2)}{h_2(x, r_1|x|^2)}\right). \quad (9.24)$$

Proof Similar to the proof of Lemma 9.7.   □

Lemma 9.8 shows that the determination of $P(x, 0)$ reduces to the determination of the solution of the following Riemann–Hilbert boundary value problem on the contour R: Determine a function $P(x, 0)$ such that

1. $P(x, 0)$ is regular for $x \in R^+$ and continuous for $x \in R \cup R^+$.

2. $\text{Re} \, (iP(x, 0)) = d(x)$, for $x \in R$,

where

$$d(x) = \text{Im} \left(P(0, 0) \frac{h_4(x, r_1|x|^2)}{h_2(x, r_1|x|^2)} \right). \tag{9.25}$$

Note that this problem is inherently different from the Riemann–Hilbert problem for $P(0, y)$ discussed in the previous section, in the sense that there is no symmetry in x and y. Moreover, the contours on which the problems have been defined have different features as well, see Lemmas 9.2 and 9.5.

We introduce the conformal mapping

$$z = g(x) : R^+ \to C^+, \tag{9.26}$$

and its inverse

$$x = g_0(z) : C^+ \to R^+, \tag{9.27}$$

which again allows us to reduce the Riemann–Hilbert problem to a Dirichlet problem on the unit circle: Determine a function $H(z)$ such that

1. $H(z)$ is regular for $z \in C^+$ and continuous for $z \in C \cup C^+$.

2. $\text{Re} \, (iH(z)) = \tilde{d}(z)$, for $z \in C$, where $\tilde{d}(z) = d(g_0(z))$.

This implies that the solution of $P(x, 0)$ is given by

$$P(x, 0) = H(g(x)) = -\frac{1}{2\pi} \oint_C d(g_0(w)) \frac{w + g(x)}{w - g(x)} \frac{dw}{w} + K_2, \quad x \in R \cup R^+, \tag{9.28}$$

where K_2 is some constant.

For the particular case that $r_1 = 1$, our contour R coincides with a contour in Blanc [17], in which a paired service model is studied using boundary value theory. In this case, an explicit expressions for the conformal mapping $g(x)$ is given by, see [17], p. 882:

$$g(x) = 1 - \frac{2\delta(1 - x)^2(1 - xr_2)}{x(1 - \delta)^2(1 - \delta r_2)} \left(1 + \frac{x - \delta}{\delta(1 - x)} \sqrt{\frac{1 - x\delta^2 r_2}{1 - xr_2}} \right), \tag{9.29}$$

where $\delta = (1 - \sqrt{1 + 8r_2})/(4r_2)$. Unfortunately, we have not been able to derive an exact expression for $g(x)$ in the case that $r_1 \neq 1$. When an explicit expression for $g(x)$ is not available, the standard approach is to determine the inverse mapping $g_0(z)$ using a well-known method from the theory of conformal mappings. This is sufficient to calculate (9.28), since we show in Sect. 9.6 that we do not need the mapping $g(x)$ to evaluate $P(x,0)$ in x.

For this approach, we need a representation of R in terms of polar coordinates, i.e.

$$R = \{x : x = \rho(\phi)\exp(i\phi),\ 0 \le \phi \le 2\pi\}, \tag{9.30}$$

which can be obtained in the following way. Since $0 \in R^+$, we have by (9.11) that for each point x on R the relation between its absolute value and its real part is given by $|x|^2 = m(\text{Re}(x))$, where

$$m(\delta) := \frac{1}{r_1 r_2(\hat{r} - 2\delta)}. \tag{9.31}$$

So, given the angle ϕ belonging to some point on R, the real part of this point, to be denoted by $\delta(\phi)$, is the solution of

$$\delta - \cos\phi\sqrt{m(\delta)} = 0, \quad 0 \le \phi \le 2\pi. \tag{9.32}$$

The question arises when the solution to (9.32) is unique. This is the case when R is a Jordan curve for which it holds that every ray from the point 0 intersects the curve R exactly once. In fact, this is the notion of starshapedness, see Pólya and Szegö [139], p. 125, Exercise 109. In all cases we have considered, R is a smooth and egg-shaped contour, and thus a starshaped Jordan curve. We see that $\rho(\phi) = \delta(\phi)/\cos\phi$, and so the parametrisation in (9.30) is fully specified.

For a contour that can be described in polar coordinates, the mapping from C^+ to the interior of this contour is formally given by, cf. Cohen and Boxma [39], Sect. I.4.4, or Gaier [69], Sect. 2.1:

$$g_0(z) = z\exp\left(\frac{1}{2\pi}\int_0^{2\pi}\log(\rho(\theta(\omega)))\frac{e^{i\omega}+z}{e^{i\omega}-z}d\omega\right), \quad |z| < 1, \tag{9.33}$$

where the angular deformation, $\theta(\cdot)$, is uniquely determined as the solution of Theodorsen's integral equation

$$\theta(\phi) = \phi - \int_0^{2\pi}\log(\rho(\theta(\omega)))\cot\left(\frac{1}{2}(\omega - \phi)\right)d\omega, \quad 0 \le \phi \le 2\pi. \tag{9.34}$$

Here, $\theta(\phi)$ is a strictly increasing and continuous function of ϕ, and $\theta(\phi) = 2\pi - \theta(2\pi - \phi)$. According to the corresponding-boundaries theorem, see Evgrafov [60], $g_0(z)$ is continuous in $C \cup C^+$. Equation (9.34) is nonlinear and cannot be solved in closed form, though a unique solution can be proven to exist.

We use (9.34) to determine boundary correspondence points. That is, for a point on the unit circle given by its angle ϕ, we solve (9.34) numerically to obtain the corresponding point on R, given by its angle $\theta(\phi)$, see Fig. 9.3. The numerical issues of this procedure are discussed in Sect. 9.6.

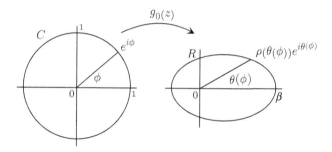

Fig. 9.3. Finding a boundary correspondence point through the mapping $x = g_0(z) : C \to R$

9.6 Performance Measures

In this section we present exact expressions for two performance measures: the fraction of time a station is empty and the mean stationary queue length at a station. Furthermore, we show for both performance measures that a relation exists between its value at station 1 and station 2. As a consequence, an expression for a performance measure for one of the stations yields the performance measure for the other station as well.

The fractions of time stations 1 and 2 are empty are given by $P(0,1)$ and $P(1,0)$, respectively. Determining either $P(0,1)$ or $P(1,0)$ is sufficient to obtain both, since they are related in the following way. Setting $x = y$ in (9.1) and taking the limit $x \uparrow 1$ gives

$$P(1,0) = 1 - \frac{\lambda}{\nu_2} - \frac{p}{1-p}\left[1 - \frac{\lambda}{\nu_1} - P(0,1)\right]. \tag{9.35}$$

Equation (9.35) alternatively follows from one of the equations

$$\lambda = \nu_1(P(1,0) - P(0,0)) + p\nu_1(1 - P(1,0) - P(0,1) + P(0,0)),$$
$$\lambda = \nu_2(P(0,1) - P(0,0)) + (1-p)\nu_2(1 - P(1,0) - P(0,1) + P(0,0)).$$

These equations stem from the following reasoning: $P(1,0) - P(0,0)$ is the fraction of time station 1 is nonempty while station 2 is empty, and $1 - P(1,0) - P(0,1) + P(0,0)$ is the fraction of time both stations are nonempty. Thus, the first equation states that, for station 1, the arrival rate equals the departure rate. Similarly, the second equation corresponds to the equality of arrival-departure rates for station 2. Note that the equations are dependent and therefore do not yield an explicit solution for $P(1,0)$ and $P(0,1)$.

We will now derive expressions for the mean queue length at both stations, to be denoted by $\mathbb{E}X_1$ and $\mathbb{E}X_2$. First, we show how these mean queue lengths are related. Differentiating both sides of (9.3) w.r.t. y, and letting $y \uparrow 1$, yields

$$\mathbb{E}(X_1)\left[\frac{1}{\nu_1} + \frac{1}{\nu_2}\right] + \mathbb{E}(X_2)\frac{1}{\nu_2} = \frac{\lambda(\nu_1^2 + \nu_1\nu_2 + \nu_2^2)}{\nu_1\nu_2(\nu_1\nu_2 - \lambda(\nu_1 + \nu_2))}. \tag{9.36}$$

Again, an interpretation can be given. The left-hand side of (9.36) measures the mean amount of work in the system by multiplying the mean number of jobs at each station by the mean service time they still require before leaving the system. The right-hand side of (9.36) corresponds to the mean amount of work in an $M/G/1$ queue, see, e.g. Cohen [35], with Poisson arrivals with rate λ and service times distributed as the sum of two independent and exponentially distributed random variables with mean $1/\nu_1$ and $1/\nu_2$, respectively. Both sides of (9.36) are equal due to the work conservation property of the system. By (9.36) it suffices to calculate either $\mathbb{E}X_1$ or $\mathbb{E}X_2$ to obtain them both. We will show how $\mathbb{E}X_2$ and $\mathbb{E}X_1$ follow from the solution of the Riemann–Hilbert boundary value problems discussed in Sects. 9.4 and 9.5, respectively.

When setting $x = 1$ in (9.1), the factor $(y-1)$ cancels from all terms leaving

$$
\begin{aligned}
P(1, y) & = \frac{\nu_1 y + \nu_2}{p\nu_1 y - (1-p)\nu_2}(-(1-p)P(1,0) + pP(0,y)) \\
& + \frac{(1-p)\nu_1 y - p\nu_2}{p\nu_1 y - (1-p)\nu_2}P(0,0). \quad (9.37)
\end{aligned}
$$

Taking derivatives w.r.t. y at both sides of (9.37) yields

$$
\begin{aligned}
\frac{d}{dy}P(1, y) & = \frac{\nu_1\nu_2}{(p\nu_1 y - (1-p)\nu_2)^2}((1-p)P(1,0) - pP(0,y)) \\
& + \frac{(\nu_1 y + \nu_2)p}{p\nu_1 y - (1-p)\nu_2}\frac{d}{dy}P(0,y) \\
& + \frac{(2p-1)\nu_1\nu_2 P(0,0)}{(p\nu_1 y - (1-p)\nu_2)^2}. \quad (9.38)
\end{aligned}
$$

Plugging (9.35) into (9.38) and setting $y = 1$ then gives for $p\nu_1 \neq (1-p)\nu_2$:

$$
\begin{aligned}
\mathbb{E}(X_2) & = \left[\frac{d}{dy}P(1,y)\right]_{y=1} \\
& = -\frac{\lambda}{p\nu_1 - (1-p)\nu_2} + \frac{(\nu_1+\nu_2)p}{p\nu_1 - (1-p)\nu_2}\left[\frac{d}{dy}P(0,y)\right]_{y=1}. \\
& \quad (9.39)
\end{aligned}
$$

Thus, to determine $\mathbb{E}X_2$, we only need to compute $[\frac{d}{dy}P(0,y)]_{y=1}$. Note that from (9.20) we have that, for $y \in L \cup L^+$,

$$
\frac{d}{dy}P(0, y) = -\frac{1}{\pi}\oint_C c(f_0(w))\frac{f'(y)}{(w - f(y))^2}dw. \quad (9.40)
$$

Similarly, when setting $y = 1$ in (9.1), the factor $(x - 1)$ cancels from all terms, which gives

$$
P(x, 1) = \frac{(1-p)\nu_1}{p\nu_1 - \lambda x}\left[\frac{p}{1-p}P(0,1) - P(x,0) + P(0,0)\right]. \quad (9.41)
$$

Taking derivatives w.r.t. x at both sides of (9.41) yields

$$\frac{d}{dx}P(x,1) = \frac{\lambda \nu_1 (1-p)}{(p\nu_1 - \lambda x)^2}\left[\frac{p}{1-p}P(0,1) - P(x,0) + P(0,0)\right]$$
$$-\frac{(1-p)\nu_1}{p\nu_1 - \lambda x}\frac{d}{dx}P(x,0). \tag{9.42}$$

Plugging (9.35) into (9.42) and setting $x = 1$ then gives for $\lambda \neq p\nu_1$:

$$\mathbb{E}(X_1) = \left[\frac{d}{dx}P(x,1)\right]_{x=1} = \frac{\lambda}{p\nu_1 - \lambda} - \frac{(1-p)\nu_1}{p\nu_1 - \lambda}\left[\frac{d}{dx}P(x,0)\right]_{x=1}. \tag{9.43}$$

Note that from (9.28) we have that, for $x \in R \cup R^+$,

$$\frac{d}{dx}P(x,0) = -\frac{1}{\pi}\oint_C d(g_0(w))\frac{g'(x)}{(w - g(x))^2}dw. \tag{9.44}$$

REMARK 9.2 When $p\nu_1 = (1-p)\nu_2$, setting $y = 1$ in (9.37) gives, after applying l'Hôpital,

$$\left[\frac{d}{dy}P(0,y)\right]_{y=1} = \frac{\lambda}{(\nu_1 + \nu_2)p} = \frac{\lambda}{\nu_2}.$$

We then have an exact expression for $[\frac{d}{dy}P(0,y)]_{y=1}$, but we cannot use (9.38) to determine $\mathbb{E}X_2$. We can, though, use (9.42) to find $\mathbb{E}X_1$, and $\mathbb{E}X_2$ through (9.36), since λ is always smaller than $p\nu_1$, when $p\nu_1 = (1-p)\nu_2$, due to the ergodicity condition (9.1). Likewise, for $\lambda = p\nu_1$, setting $x = 1$ in (9.41) gives, after applying l'Hôpital,

$$\left[\frac{d}{dx}P(x,0)\right]_{x=1} = \frac{\lambda}{(1-p)\nu_1} = \frac{\lambda}{\nu_1 - \lambda},$$

and we cannot use (9.42) to determine $\mathbb{E}X_1$. We can use (9.38) to find $\mathbb{E}X_2$, since $(1-p)\nu_2$ is always smaller than $p\nu_1$ (when $\lambda = p\nu_1$) due to the ergodicity condition. We can thus conclude that we can calculate either one of the integrals (9.40) or (9.44) for all allowed parameter values.

Let us now discuss some issues that arise in computing the performance measures from the formal solutions of the Riemann–Hilbert boundary value problems. In Sect. 9.6.1 we discuss how the location of α and β is related to the set of parameter values for which we can actually determine the performance measures. In Sect. 9.6.2 we discuss a way to determine the performance measures for all allowed parameter values. In Sect. 9.6.3 we discuss how the integrals involved in computing the performance measures can be determined numerically. Finally, we present some conclusions in Sect. 9.6.4.

9.6.1 Remarks on α and β

For calculating the performance measures described in Sect. 9.6, we have to evaluate $P(0,y)$ and $\frac{d}{dy}P(0,y)$ in $y = 1$ or $P(x,0)$ and $\frac{d}{dx}P(x,0)$ in $x = 1$. We first discuss the first option. The integration constant K_1 can be determined by calculating $P(0,0)$ from the integral (9.20), and using that $P(0,0) = 1 - \lambda/\nu_1 - \lambda/\nu_2$. The integrals (9.20), (9.40), however, follow from the solution of a Dirichlet problem that is only defined on or within the unit circle. So, in order to evaluate the integrals, $f(1)$ should lie on or within the unit circle, which is the same as requiring 1 to lie on or within the contour L.

The above problem is very common in queueing applications for which the boundary value technique is applied, see, e.g. Boxma and Groenendijk [24], Cohen and Boxma [39], p. 360, De Klein [96], p. 89, Feng et al. [66], Mikou [123], and Mikou et al. [124]. In the present context, a key role is played by α. In Lemma 9.3 we saw that, when $p\nu_1 \le (1 - p)\nu_2$ (i.e. $r_1 \ge r_2$), it follows that $\alpha \ge 1$ and thus $1 \in L \cup L^+$. Hence, for these parameter values the integrals (9.20), (9.40) can be calculated. To obtain results for parameter values for which it holds that $p\nu_1 > (1 - p)\nu_2$, we might consider analytic continuations for the functions (9.20), (9.40), see, e.g. Nauta [130]. However, this would most probably result in numerical difficulties. Alternatively, we can use the Taylor series expansions of the corresponding functions around some point in L^+, as suggested by Cohen and Boxma [39], p. 360. For a Taylor series expansion of order n around $\hat{y} \in L^+$, we then have that

$$P(0,1) \approx \sum_{k=0}^{n} \frac{(1 - \hat{y})^k}{k!} \left[\frac{d^k}{dy^k} P(0,y) \right]_{y=\hat{y}}. \tag{9.45}$$

The exact same problem applies for boundary value problem II. In that case, we have to evaluate the integrals (9.28), (9.44) in $x = 1$, which is only allowed when $1 \in R \cup R^+$. In Lemma 9.6 we have seen that this is the case when $p\nu_1$ and $(1 - p)\nu_2$ are both larger than λ (i.e. $r_1 < 1$ and $r_2 < 1$).

To summarise, we show in Fig. 9.4 how the values of α and β are related to the parameter values λ, ν_1, ν_2 (for $p = 1/2$). So, starting from boundary value problem I, we can determine the performance measures for parameter values that fall within areas I and II. By considering boundary value problem II, we can enlarge this set by area III. For area IV we can apply the Taylor series expansion. As will be shown in the next section, the use of Taylor series expansions can be circumvented by considering a third zero-set of the kernel $h_1(x,y)$.

9.6.2 A Third Zero-Set of the Kernel

We now discuss an approach to determine $P(0,1)$ and $[\frac{d}{dy}P(0,y)]_{y=1}$ directly from (9.20), (9.40) despite the fact that $\alpha < 1$. The approach has been

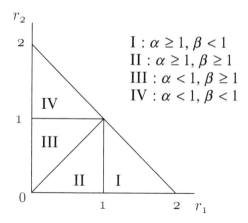

Fig. 9.4. Relation between parameter values, and α and β, for $p = \frac{1}{2}$

suggested by De Klein [96], p. 89, and makes use of a zero-set of $h_1(x, y)$ other than the ones we have considered so far. By establishing a relation between $P(x, 0)$ and $P(0, y)$ for zero-pairs (x, y) of this set, we are able to calculate the performance measures for all allowed parameter values.

The new zero-set is defined by

$$\{(x, y^*(x)) \mid h_1(x, y^*(x)) = 0, |x| = 1\}, \tag{9.46}$$

where $y^*(x)$ is the zero of the kernel with the smallest modulus. From the function $y(x)$, as given in (9.5), it is easily seen that

$$y^*(1) = \min\left(\frac{r_1}{r_2}, 1\right), \tag{9.47}$$

for which we have the following result:

LEMMA 9.9 *For $r_1 = r_2$ it holds that $y^*(1) = 1 = \alpha$. For $r_1 \neq r_2$ it holds that $y^*(1) < \alpha$.*

Proof The first assertion follows immediately from Lemma 9.3. For the second assertion note that if $r_1 > r_2$ it holds that $y^*(1) = 1 < \alpha$. If $r_1 < r_2$ it holds that

$$D_1(r_1/r_2) = \left[1 + r_1 + \frac{r_1}{r_2} - \frac{r_1^2}{r_2}\right]^2 \left(\frac{1}{r_2}\right)^2 - 4\left(\frac{1}{r_2}\right)^2.$$

Since $r_1 < r_2$ implies $r_1 < 1$, we have $1 + r_1 + r_1/r_2 - r_1^2/r_2 \in (-2, 2)$, and $D_1(r_1/r_2) < 0$. So, $r_1/r_2 < x_2$, and $y^*(1) = r_1/r_2 < \sqrt{r_1 x_2/r_2} = \alpha$. □

We exploit the result in Lemma 9.9 in the following way. Introducing the short-hand notation $h_k(x) := h_k(x, y^*(x))$, we obtain from (9.1) that

$$h_2(x)P(x, 0) + h_3(x)P(0, y^*(x)) + h_4(x)P(0, 0) = 0, \quad |x| = 1. \tag{9.48}$$

Setting $x = 1$ in (9.48) yields

$$P(1,0) = \frac{-1}{h_2(1)}\Big[h_3(1)P(0, y^*(1)) + h_4(1)P(0,0)\Big]. \qquad (9.49)$$

Since for $r_1 \neq r_2$ it holds that $y^*(1) < \alpha$, the value of $P(0, y^*(1))$ can be computed directly from (9.20). Hence, for $r_1 < r_2$ we cannot obtain $P(0,1)$ directly from (9.20), but we can obtain $P(1,0)$ using (9.49), and find $P(0,1)$ through (9.35).

By using a similar approach we can determine $[\frac{d}{dy}P(0, y)]_{y=1}$ through (9.40), despite the fact that $r_1 < r_2$. We do need some extra results concerning the zero-set (9.46) though. Observe that $y^*(1)$ is of multiplicity 1 unless $r_1 = r_2$, for which $y^*(1)$ is of multiplicity two. We further have

LEMMA 9.10 *The zero $y^*(x)$ is of multiplicity 1 and contained in the disk $|y| < 1$ for every $|x| = 1$, $x \neq 1$.*

Proof For $|x| = 1$ it holds that $h_1(x, y) = \lambda x(f(x, y) + g(x, y))$ where

$$f(x, y) := \Big(1 + \frac{1}{r_1} + \frac{1}{r_2} - x\Big)y, \quad g(x, y) := -\Big(\frac{1}{r_1}\bar{x}y^2 + \frac{1}{r_2}\Big),$$

and $\bar{x}$ the complex conjugate of x. We have for $|x| = 1$, $x \neq 1$,

$$|f(x, y)| = |1 + \frac{1}{r_1} + \frac{1}{r_2} - x||y| > \Big(\frac{1}{r_1} + \frac{1}{r_2}\Big)|y|,$$

$$|g(x, y)| \leq \frac{1}{r_1}|\bar{x}||y|^2 + \frac{1}{r_2} = \frac{1}{r_1}|y|^2 + \frac{1}{r_2}.$$

Then, for all points y on $|y| = 1$ we have that

$$|g(x, y)| \leq \frac{1}{r_1} + \frac{1}{r_2} < |f(x, y)|, \quad |y| = 1, |x| = 1, x \neq 1,$$

which implies by Rouché's theorem, see, e.g. Titchmarsh [162], that $f(x, y) + g(x, y)$ (and thus $h_1(x, y)$) has as many zeros, counted according to their multiplicity, inside $|y| = 1$ as $f(x, y)$. Since $f(x, y)$ has only one zero of multiplicity 1 at $y = 0$, we find that for every x with $|x| = 1$, $x \neq 1$, $h_1(x, y) = 0$ has one solution inside $|y| = 1$, i.e. $y^*(x)$. □

From Lemma 9.10 it follows for $r_1 \neq r_2$ and $|x| = 1$ that

$$\Big[\frac{d}{dy}h_1(x, y)\Big]_{y=y^*(x)} \neq 0, \qquad (9.50)$$

because otherwise $y^*(x)$ would be of multiplicity 2. From the implicit function theorem we then have that $y^*(x)$ is differentiable for $r_1 \neq r_2$ and $|x| = 1$.

Differentiating $h_1(x, y^*(x)) = 0$ at both sides gives

$$\left[\frac{\mathrm{d}}{\mathrm{d}x}h_1(x, y)\right]_{y=y^*(x)} + \frac{\mathrm{d}}{\mathrm{d}x}y^*(x)\left[\frac{\mathrm{d}}{\mathrm{d}y}h_1(x, y)\right]_{y=y^*(x)} = 0, \qquad (9.51)$$

and thus

$$\frac{\mathrm{d}}{\mathrm{d}x}y^*(x) = -\frac{\left[\frac{\mathrm{d}}{\mathrm{d}x}h_1(x, y)\right]_{y=y^*(x)}}{\left[\frac{\mathrm{d}}{\mathrm{d}y}h_1(x, y)\right]_{y=y^*(x)}}. \qquad (9.52)$$

Consequently, differentiating (9.48) w.r.t. x and setting $x = 1$ gives

$$\left[\frac{\mathrm{d}}{\mathrm{d}x}P(x, 0)\right]_{x=1} = \frac{-1}{h_2(1)}\left(h_2'(1)P(1, 0) + h_3'(1)P(0, y^*(1))\right)$$

$$+ h_4'(1)P(0, 0) + h_3(1)\left[\frac{\mathrm{d}}{\mathrm{d}x}y^*(x)\right]_{x=1}\left[\frac{\mathrm{d}}{\mathrm{d}y}P(0, y)\right]_{y=y^*(1)}\right). \qquad (9.53)$$

Again, since for $r_1 \neq r_2$ it holds that $y^*(1) < \alpha$, the value of $\frac{\mathrm{d}}{\mathrm{d}y}P(0, y)$ in $y = y^*(1)$ can be computed directly from (9.40), and through (9.53), (9.42), (9.36) and (9.38) we obtain $(\mathrm{d}/\mathrm{d}y)P(0, y)$ in $y = 1$. The approach outlined in this section can also be applied to determine $P(1, 0)$ and $(\mathrm{d}/\mathrm{d}x)P(x, 0)$ in $x = 1$ in case $\beta < 1$.

9.6.3 Evaluating the Integrals

We will now describe how the involved integrals can be determined numerically. For boundary value problem I, we have rewritten the integral (9.20) as (9.21). The integral (9.40) can be rewritten in a similar way. We will evaluate the integrals (9.21) and (9.40) using the trapezium rule, for which we split the interval $[0, 2\pi]$ into K parts of equal length $2\pi/K$. The fact that the whole integrand including the mapping $f(y)$ is known explicitly allows for a fine subdivision. For the numerical results to be presented in the next section we have set K to 250, which guarantees a high level of accuracy.

For boundary value problem II, we need to calculate the integrals (9.28) and (9.44). We will now outline how these integrals can be computed, along with the numerical determination of the mapping $g_0(z)$. For a more detailed exposition we refer to Chap. IV.1 of Cohen and Boxma [39].

Step 1: Rewriting the integrals (9.28) *and* (9.44)
Substitution of $w = e^{i\phi}$ into (9.28) yields

$$P(x, 0) = -\frac{i}{2\pi}\int_0^{2\pi} d(g_0(e^{i\phi}))\frac{e^{i\phi} + g(x)}{e^{i\phi} - g(x)}\mathrm{d}\phi + K_2, \qquad x \in R \cup R^+. \qquad (9.54)$$

The integral (9.44) can be rewritten in a similar way, i.e.

$$\frac{\mathrm{d}}{\mathrm{d}x}P(x,0) = -\frac{i}{\pi}\int_0^{2\pi} d(g_0(e^{i\phi}))\frac{g'(x)e^{i\phi}}{(e^{i\phi}-g(x))^2}\mathrm{d}\phi, \quad x \in R \cup R^+. \quad (9.55)$$

Step 2: Numerical evaluation of the integrals (9.54) and (9.55)
We will evaluate the integrals (9.54) and (9.55) in $x = 1$ using the above rewriting and the trapezium rule, for which we split the interval $[0, 2\pi]$ into K parts of equal length $2\pi/K$. From (9.54) and (9.55) we then see that we need to determine the values of the conformal mapping $g_0(\cdot)$ in the points $e^{i\phi_k}$, $k = 0, 1, \ldots, K - 1$, with $\phi_k = 2\pi k/K$. We further need to determine $g(1)$ and $g'(1)$.

Step 3: Solving Theodorsen's integral equation (9.34)
For K points on the unit circle given by their angles

$$\{\phi_0, \phi_1, \ldots, \phi_{K-1}\},$$

we need to solve (9.34) to obtain the corresponding points on R, given by their angles $\{\theta(\phi_0), \theta(\phi_1), \ldots, \theta(\phi_{K-1})\}$. For $k = 0, 1, \ldots, K - 1$, we determine $\theta(\phi_k)$ iteratively, see Gaier [69], p. 67, from

$$\theta_0(\phi_k) = \phi_k, \quad (9.56)$$

$$\theta_{n+1}(\phi_k) = \phi_k - \int_0^{2\pi} \log\left(\frac{\delta(\theta_n(\omega))}{\cos(\theta_n(\omega))}\right)\cot\left(\frac{1}{2}(\omega - \phi_k)\right)\mathrm{d}\omega, \quad (9.57)$$

where $\delta(\theta_n(\omega))$ is determined from, see (9.32),

$$\delta(\theta_n(\omega)) - \cos\theta_n(\omega)\sqrt{m(\delta(\theta_n(\omega)))} = 0, \quad (9.58)$$

using the Newton-Raphson root-finding procedure. For each step, the integral in (9.57) is numerically determined by again using the trapezium rule with K parts of equal length $2\pi/K$. For the iteration, we have used the following stopping criterion:

$$\max_{k\in\{0,\ldots,K-1\}} |\theta_{n+1}(\phi_k) - \theta_n(\phi_k)| < 10^{-6}. \quad (9.59)$$

Finally, it follows from $\rho(\phi) = \delta(\phi)/\cos\phi$ that the value of $g_0(\cdot)$ in $e^{i\phi_k}$ is given by

$$g_0(e^{i\phi_k}) = \frac{\delta(\theta(\phi_k))}{\cos\theta(\phi_k)}e^{i\theta(\phi_k)}, \quad k = 0, 1, \ldots, K - 1. \quad (9.60)$$

We again set K to 250, although in our experience a far smaller value of K is already sufficient to reach an acceptable level of accuracy.

Step 4: Determination of $g(1)$ and $g'(1)$

$g(1)$ is obtained as the unique solution z of $g_0(z) = 1$ on $[0, 1]$, and can be determined using (9.33) and Newton-Raphson. $g'(1)$ is given by, see Boxma and Groenendijk [24],

$$g'(1) = \left[\frac{1}{g(1)} + \frac{1}{2\pi} \exp\left(\frac{1}{2\pi} \int_0^{2\pi} \log\left(\frac{\delta(\theta(\omega))}{\cos(\theta(\omega))} \right) \frac{2e^{i\omega}}{(e^{i\omega} - g(1))^2} d\omega \right) \right]^{-1}.$$

(9.61)

We calculate $g'(1)$ by numerically determining the integral (9.60) with the trapezium rule and K set to 250.

9.6.4 Conclusion

For both boundary value problems, determining the performance measures comes down to computing real integrals. For boundary value problem I, we have an explicit expression for the conformal mapping $f(y)$, and so computing the real integrals becomes a standard exercise. For boundary value problem II, though, we are not able to derive the required conformal mapping $g(x)$. We therefore choose to numerically determine its inverse conformal mapping in order to compute the integrals. Hence, using boundary value problem II requires some additional effort.

In Sect. 9.6.1 we saw that both models were useful in computing the performance measures. However, using the approach outlined in Sect. 9.6.2, boundary value problem I can be applied to determine the performance measures for the complete range of allowed parameter values. Hence, we naturally suggest to use boundary value problem I for computational purposes. If, however, one could derive an explicit expression for the mapping $g(x)$, boundary value problem II would be equally suitable.

9.6.5 Some Examples

In this section we present some examples that show the effect of the value of the parameter p on the performance measures. In Sect. 9.6 we have concluded that we can determine the performance measures for the whole set of parameter values $\{\lambda, \nu_1, \nu_2, p\}$ for which the ergodicity condition (9.1) is satisfied. Moreover, we have seen that part of this set allows for multiple ways to determine the performance measures. Therefore, we cross-checked all results presented in this section whenever possible.

Table 9.1 displays the performance measures for a moderate load ($P(0,0) = 1/3$). The results for the limiting cases $p = 0$ and $p = 1$ are obtained with the solutions as given in Resing and Örmeci [148]. Obvious observations are that the fraction of time station 1 is busy, and the mean queue length at station 1, both decrease for higher values of p, and vice versa for station 2. Further

Table 9.1. Performance measures for moderate load ($P(0,0) = 1/3$)

λ	ν_1	ν_2	p	r_1	r_2	α	β	$P(1,0)$	$P(0,1)$	$\mathbb{E}X_1$	$\mathbb{E}X_2$
1	3	3	0.00	∞	0.33	–	–	0.44	0.67	1.33	0.33
1	3	3	0.25	1.33	0.44	1.32	0.86	0.47	0.60	1.25	0.50
1	3	3	0.50	0.67	0.67	1.00	1.15	0.52	0.52	1.02	0.97
1	3	3	0.75	0.44	1.33	0.44	0.86	0.60	0.47	0.68	1.63
1	3	3	1.00	0.33	∞	–	–	0.67	0.44	0.50	2.00
1	6	2	0.00	∞	0.50	–	–	0.50	0.50	1.25	0.50
1	6	2	0.25	0.67	0.67	1.00	1.15	0.61	0.43	0.76	1.16
1	6	2	0.50	0.33	1.00	0.46	1.00	0.73	0.40	0.38	1.67
1	6	2	0.75	0.22	2.00	0.17	0.70	0.80	0.39	0.26	1.83
1	6	2	1.00	0.17	∞	–	–	0.83	0.39	0.20	1.90
1	2	6	0.00	∞	0.17	–	–	0.39	0.83	1.58	0.17
1	2	6	0.25	2.00	0.22	1.57	0.70	0.39	0.80	1.57	0.22
1	2	6	0.50	1.00	0.33	1.37	1.00	0.40	0.73	1.54	0.34
1	2	6	0.75	0.67	0.67	1.00	1.15	0.43	0.61	1.42	0.83
1	2	6	1.00	0.50	∞	–	–	0.50	0.50	1.00	2.50

Table 9.2. Performance measures for high load ($P(0,0) = 0.1$)

λ	ν_1	ν_2	p	r_1	r_2	α	β	$P(1,0)$	$P(0,1)$	$\mathbb{E}X_1$	$\mathbb{E}X_2$
1.8	4	4	0.00	∞	0.45	–	–	0.15	0.55	6.53	0.45
1.8	4	4	0.25	1.80	0.60	1.22	0.75	0.16	0.42	6.34	0.82
1.8	4	4	0.50	0.90	0.90	1.00	1.04	0.22	0.22	4.63	4.23
1.8	4	4	0.75	0.60	1.80	0.41	0.75	0.42	0.16	1.42	10.66
1.8	4	4	1.00	0.45	∞	–	–	0.55	0.15	0.82	11.86
1.8	6	3	0.00	∞	0.60	–	–	0.16	0.40	6.60	0.60
1.8	6	3	0.25	1.20	0.80	1.09	0.90	0.20	0.23	5.78	1.83
1.8	6	3	0.50	0.60	1.20	0.60	0.90	0.45	0.15	1.31	8.53
1.8	6	3	0.75	0.40	2.40	0.23	0.65	0.61	0.13	0.64	9.54
1.8	6	3	1.00	0.30	∞	–	–	0.70	0.13	0.43	9.86
1.8	3	6	0.00	∞	0.30	–	–	0.13	0.70	6.90	0.30
1.8	3	6	0.25	2.40	0.40	1.38	0.65	0.13	0.61	6.85	0.82
1.8	3	6	0.50	1.20	0.60	1.19	0.90	0.15	0.45	6.67	4.23
1.8	3	6	0.75	0.80	1.20	0.73	0.90	0.23	0.20	3.57	10.26
1.8	3	6	1.00	0.60	∞	–	–	0.40	0.16	1.50	16.38

note that $\mathbb{E}X_1 + \mathbb{E}X_2$ increases as a function of p. Table 9.2 displays the performance measures for a high load ($P(0,0) = 0.1$), from which a similar conclusion can be drawn.

For procedures that require two sequential stages, balancing either the mean queue length or the mean workload might be of interest, see, e.g. Andradottir et al. [6]. In Tables 9.1 and 9.2 we see that the difference in mean queue lengths at the two stations is strongly influenced by p. As an example, we have plotted

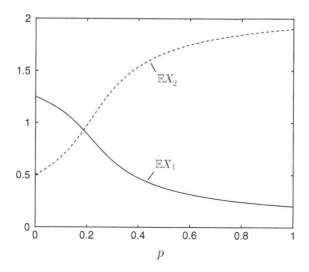

Fig. 9.5. Mean queue lengths for $\lambda = 1, \nu_1 = 6, \nu_2 = 2$, and $p \in [0, 1]$

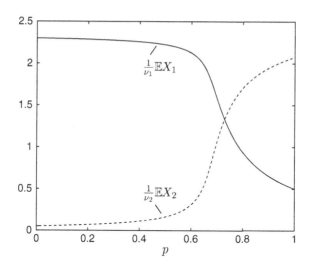

Fig. 9.6. Mean workloads for $\lambda = 1.8, \nu_1 = 3, \nu_2 = 6$, and $p \in [0, 1]$

in Fig. 9.5 both mean queue lengths for $\lambda = 1, \nu_1 = 6, \nu_2 = 2$, and p running from 0 to 1. The imbalance is minimal when $\mathbb{E}X_1 = \mathbb{E}X_2$, i.e. $p \approx 0.18$. Observe that the optimal value of p does not correspond to the solution of $p\nu_1 = (1 - p)\nu_2$, which is 0.25.

An example of the influence of p on the mean workloads $\frac{1}{\nu_1}\mathbb{E}X_1$ and $\frac{1}{\nu_2}\mathbb{E}X_2$ is given by Fig. 9.6, which shows the mean workloads for $\lambda = 1.8, \nu_1 = 3, \nu_2 = 6$, and p running from 0 to 1. We observe that the imbalance in workloads is minimal when $\frac{1}{\nu_1}\mathbb{E}X_1 = \frac{1}{\nu_2}\mathbb{E}X_2$, i.e. $p \approx 0.72$.

Chapter 10

A TWO-STATION NETWORK
WITH COUPLED PROCESSORS

In Chap. 9 we have considered the two-stage tandem queue with coupled processors. We have shown that the pgf of the joint stationary queue length distribution can be found using the theory of boundary value problems.

In this chapter we present a more general model of a two-station network with coupled processors. This network consists of two single-server stations, Poisson arrival streams, exponential service times and probabilistic routing. For the network with coupled processors, we will show that the pgf of the joint stationary queue length distribution can again be found using the theory of boundary value problems. The network provides a general framework which, among other things, covers and hence generalises two key models: The two-stage tandem queue and two parallel queues. The second model has been studied by Fayolle and Iasnogorodski [61]. Their paper has become a classic one, since it introduced the technique of boundary value problems to the field of queueing theory.

10.1 Introduction

In recent years, the coupled-processors discipline has regained attention. This is due to the fact that the coupled-processors discipline incorporates the generalised processor sharing discipline (GPS) as a special case. GPS is a popular scheduling discipline in modern communication networks, since it provides a way to achieve service differentiation among different types of customers. The recent work on GPS is focused on deriving characteristics of the queue length and delay distributions, particularly on deriving the asymptotic behaviour of tail probabilities, see Van Uitert [167] and the references therein, and less on deriving a transform solution of the stationary distribution, as Fayolle and Iasnogorodski [61] did. It is widely recognised that obtaining transform solutions for GPS, or coupled-processors, models with more than two customer classes is extremely hard. Cohen [36] obtained a partial solution for a three-class GPS model, but up to this day that seems to be as far as it can be taken. The two-station network studied in this chapter is in fact a description of a collection of two-dimensional models with exponential service times for which an analytical solution can be obtained in terms of a transform.

The remainder of this chapter is structured as follows. In Sect. 10.2 we give the model description and derive a functional equation for the joint pgf of the stationary queue length distribution. In Sect. 10.3 we derive expressions and relations for various performance measures. In Sect. 10.4 we treat the case of preemptive priority for one of the stations. We show that from the functional equation the joint pgf of the stationary queue length can be obtained without invoking the theory of boundary value problems. For the GPS case, we do need the theory of boundary value problems, as presented in Sect. 10.5. We end this chapter with some conclusions and suggestions for further research in Sect. 10.6.

10.2   Model Description

Consider an open queueing network with two single-server stations, where jobs arrive externally at station j according to a Poisson process with rate λ_j, $j = 1, 2$. Every time a job visits station j, $j = 1, 2$, it requires an exponential amount of work with parameter ν_j. The total service capacity of the two stations together is constant. When both stations are nonempty, the service capacity is divided between the two stations according to fixed proportions, p_1 and p_2, $p_1 + p_2 = 1$, and hence the departure rates at station 1 and 2 then equal $p_1\nu_1$ and $p_2\nu_2$, respectively. If one of the stations is empty, the total service capacity of the stations is allocated to the nonempty station. Hence, the departure rate at that station, station j say, is then temporarily increased to ν_j.

Figure 10.1 displays the network. We denote by r_{ij} the probability that a job moves to station j after receiving service at station i. After receiving service at station 1, a job will join the queue of station 2 w.p. r_{12}, or leave the system w.p. $1 - r_{12}$. Similarly, after receiving service at station 2, a job will join the queue

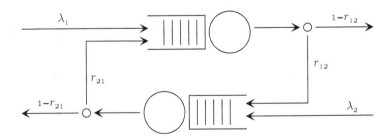

Fig. 10.1. Two-station network with probabilistic routing and coupled processors

of station 1 w.p. r_{21}, or leave the system w.p. $1 - r_{21}$. We assume that $r_{11} = 0$ and $r_{22} = 0$, which can be done without loss of generality when modelling the queue lengths, since $r_{ii} > 0$ implies that the service time of a job at station i is a geometrically distributed sum of exponentially distributed random variables, which is again exponentially distributed. Note that the network model reduces to the tandem queue covered in Chap. 9 by choosing $\lambda_2 = r_{21} = 0$ and $r_{12} = 1$.

Denoting by γ_j the throughput of jobs at station j, we have that $\gamma_1 = \lambda_1 + \gamma_2 r_{21}$ and $\gamma_2 = \lambda_2 + \gamma_1 r_{12}$, which gives

$$\gamma_1 = \frac{\lambda_1 + r_{21}\lambda_2}{1 - r_{12}r_{21}}, \quad \gamma_2 = \frac{\lambda_2 + r_{12}\lambda_1}{1 - r_{12}r_{21}}. \tag{10.1}$$

Denote by $X_j(t)$ the number of jobs at station j at time t. Under the condition,

$$\rho_1 + \rho_2 < 1, \tag{10.2}$$

where $\rho_j = \gamma_j/\nu_j$, the two-dimensional Markov process

$$\{(X_1(t), X_2(t)), t \geq 0\} \tag{10.3}$$

has a unique stationary distribution. This can be explained by the fact that, independent of p_1 and p_2, the two stations together always work at capacity 1, if there is work in the system, and that $\rho_1 + \rho_2$ equals the amount of work brought into the system per time unit. Note that in case the servers are not coupled, we have a standard two-station open Jackson network, for which the stationary joint queue length distribution possesses the following product form

$$\lim_{t \to \infty} \mathbb{P}(X_1(t) = n_1, X_2(t) = n_2) = (1 - \rho_1)\rho_1^{n_1}(1 - \rho_2)\rho_2^{n_2}. \tag{10.4}$$

For the case with coupled processors, such a product form fails to hold.

Let us denote by $\pi(n, k)$ the stationary probability of having n jobs at station 1 and k jobs at station 2. The following set of balance equations can then be derived, with $\lambda := \lambda_1 + \lambda_2$:

$$
\begin{aligned}
\lambda\pi(0,0) &= \nu_2(1 - r_{21})\pi(0,1) + \nu_1(1 - r_{12})\pi(1,0), \\
(\lambda + \nu_1)\pi(1,0) &= \lambda_1\pi(0,0) + \nu_1(1 - r_{12})\pi(2,0) + p_2\nu_2(1 - r_{21})\pi(1,1) \\
&\quad + \nu_2 r_{21}\pi(0,1), \\
(\lambda + \nu_2)\pi(0,1) &= \lambda_2\pi(0,0) + \nu_2(1 - r_{21})\pi(0,2) + p_1\nu_1(1 - r_{12})\pi(1,1) \\
&\quad + \nu_1 r_{12}\pi(1,0),
\end{aligned}
$$

and for $n \geq 1, k \geq 1$

$$
\begin{aligned}
(\lambda + \nu_1)\pi(n,0) &= \lambda_1\pi(n-1,0) + \nu_1(1 - r_{12})\pi(n+1,0) \\
&\quad + p_2\nu_2(1 - r_{21})\pi(n,1) + p_2\nu_2 r_{21}\pi(n-1,1), \\
(\lambda + \nu_2)\pi(0,k) &= \lambda_2\pi(0,k-1) + \nu_2(1 - r_{21})\pi(0,k+1) \\
&\quad + p_1\nu_1(1 - r_{12})\pi(1,k) + p_1\nu_1 r_{12}\pi(1,k-1),
\end{aligned}
$$

$$(\lambda + p_1\nu_1 + p_2\nu_2)\pi(1,1) = \lambda_1\pi(0,1) + \lambda_2\pi(1,0) + \nu_1 r_{12}\pi(2,0)$$
$$+\nu_2 r_{21}\pi(0,2) + p_1\nu_1(1 - r_{12})\pi(2,1)$$
$$+p_2\nu_2(1 - r_{21})\pi(1,2),$$

$$(\lambda + p_1\nu_1 + p_2\nu_2)\pi(n,1) = \lambda_1\pi(n-1,1) + \lambda_2\pi(n,0)$$
$$+\nu_1 r_{12}\pi(n+1,0) + p_2\nu_2 r_{21}\pi(n-1,2)$$
$$+p_1\nu_1(1 - r_{12})\pi(n+1,1)$$
$$+p_2\nu_2(1 - r_{21})\pi(n,2),$$

$$(\lambda + p_1\nu_1 + p_2\nu_2)\pi(1,k) = \lambda_2\pi(1,k-1) + \lambda_1\pi(0,k)$$
$$+\nu_2 r_{21}\pi(0,k+1) + p_1\nu_1 r_{12}\pi(2,k-1)$$
$$+p_2\nu_2(1 - r_{12})\pi(1,k+1)$$
$$+p_1\nu_1(1 - r_{12})\pi(2,k).$$

We define the joint probability generating function

$$P(x,y) = \sum_{n\geq 0}\sum_{k\geq 0}\pi(n,k)x^n y^k, \quad |x| \leq 1, \ |y| \leq 1,$$

which is regular for $|x| < 1$, continuous for $|x| \leq 1$ for every fixed y, and similarly for x and y interchanged. From the balance equations it follows that $P(x,y)$ satisfies the following functional equation

$$h_1(x,y)P(x,y) = h_2(x,y)P(x,0) + h_3(x,y)P(0,y) + h_4(x,y)P(0,0), \tag{10.5}$$

where

$$h_1(x,y) = (\lambda + p_1\nu_1 + p_2\nu_2)xy - \lambda_1 x^2 y - \lambda_2 xy^2 - p_1\nu_1 r_{12}y^2$$
$$-p_2\nu_2 r_{21}x^2 - p_1\nu_1(1 - r_{12})y - p_2\nu_2(1 - r_{21})x,$$

$$h_2(x,y) = p_2[(\nu_2 - \nu_1)xy + \nu_1 r_{12}y^2 - \nu_2 r_{21}x^2$$
$$+\nu_1(1 - r_{12})y - \nu_2(1 - r_{21})x],$$

$$h_3(x,y) = -p_1[(\nu_2 - \nu_1)xy + \nu_1 r_{12}y^2 - \nu_2 r_{21}x^2$$
$$+\nu_1(1 - r_{12})y - \nu_2(1 - r_{21})x],$$

$$h_4(x,y) = p_2\nu_1(xy - r_{12}y^2 - (1 - r_{12})y)$$
$$+p_1\nu_2(xy - r_{21}x^2 - (1 - r_{21})x).$$

Observe that the four functions above are all polynomials of degree 2 in both x and y, and that $h_2(x,y) = -(p_2/p_1)h_3(x,y)$.

10.3 Performance Measures

In this chapter we derive expressions for two performance measures: The fraction of time the stations are empty and the mean stationary queue length at the stations. The fractions of time stations 1 and 2 are empty, given by $P(0,1)$ and $P(1,0)$ respectively, are related as

$$
\begin{aligned}
\gamma_1 &= \nu_1(P(1,0) - P(0,0)) + p_1\nu_1(1 - P(1,0) - P(0,1) + P(0,0)), \\
\gamma_2 &= \nu_2(P(0,1) - P(0,0)) + p_2\nu_2(1 - P(1,0) - P(0,1) + P(0,0)).
\end{aligned}
$$

These equations stem from the following reasoning, similar to the reasoning on p. 201: $P(1,0) - P(0,0)$ is the fraction of time station 1 is nonempty while station 2 is empty, and $1 - P(1,0) - P(0,1) + P(0,0)$ is the fraction of time both stations are nonempty. Thus, the first equation states that, for station 1, the arrival rate equals the departure rate. Similarly, the second equation corresponds to the equality of arrival-departure rates for station 2. Note that the equations are dependent and therefore do not yield an explicit solution for $P(1,0)$ and $P(0,1)$.

The mean stationary queue lengths at both stations, to be denoted by $\mathbb{E}X_1$ and $\mathbb{E}X_2$, are also related. First, we give the following definition:

DEFINITION 10.1 We define by $B_{[\pi_1;\pi_2]}$ a random variable having a two-phase phase-type distribution, starting from phase 1 or phase 2 w.p. π_1 and π_2, respectively. The transition rate out of phase 1 (2) equals ν_1 (ν_2). After completing phase 1 (2), the process may continue with phase 2 (1) w.p. r_{12} (r_{21}), or enters the absorbing state.

The moments of $B_{[\pi_1;\pi_2]}$ are given by, see Asmussen [7], Proposition 4.1, p. 83,

$$
\mathbb{E}(B_{[\pi_1;\pi_2]}^k) = (-1)^k k! \begin{bmatrix} \pi_1 & \pi_2 \end{bmatrix} \begin{bmatrix} -\nu_1 & \nu_1 r_{12} \\ \nu_2 r_{21} & -\nu_2 \end{bmatrix}^{-k} \begin{bmatrix} 1 \\ 1 \end{bmatrix}. \tag{10.6}
$$

We then have the following result:

LEMMA 10.1 *The mean queue lengths at both stations are related in the following way:*

$$
\mathbb{E}X_1 \mathbb{E}(B_{[1;0]}) + \mathbb{E}X_2 \mathbb{E}(B_{[0;1]}) = \frac{\rho}{1-\rho} [2\mathbb{E}(B_{[\lambda_1/\lambda;\lambda_2/\lambda]})]^{-1} \mathbb{E}(B_{[\lambda_1/\lambda;\lambda_2/\lambda]}^2), \tag{10.7}
$$

where $\rho = \lambda \mathbb{E}(B_{[\lambda_1/\lambda;\lambda_2/\lambda]})$.

Proof The left-hand side of (10.7) counts the mean amount of work in the system by multiplying the mean number of jobs by the mean service time they still require before leaving the system. The right-hand side of (10.7) corresponds to the mean amount of work in an $M/G/1$ queue with Poisson arrivals with rate λ and service times distributed as $B_{[\lambda_1/\lambda;\lambda_2/\lambda]}$, see Cohen [35], p. 256.   □

By (10.7) it suffices to calculate either $\mathbb{E}X_1$ or $\mathbb{E}X_2$ to obtain them both. We will show how $\mathbb{E}X_2$ follows from the solution of the Riemann–Hilbert boundary value problem discussed in Sect. 10.5.

When setting $x = 1$ in (10.4), we can divide both sides by $(y-1)$, and after rewriting we obtain

$$
\begin{aligned}
P(1,y) &= \frac{\nu_2 + \nu_1 r_{12} y}{p_2\nu_2 - (\lambda_2 + p_1\nu_1 r_{12})y}(p_2 P(1,0) - p_1 P(0,y)) \\
&+ \frac{p_1\nu_2 - p_2\nu_1 r_{12}y}{p_2\nu_2 - (\lambda_2 + p_1\nu_1 r_{12})y}P(0,0).
\end{aligned}
\tag{10.8}
$$

Differentiating (10.8) w.r.t. y yields

$$
\begin{aligned}
\frac{\mathrm{d}}{\mathrm{d}y}P(1,y) &= \frac{\nu_1\nu_2 r_{12} + \nu_2\lambda_2}{(p_2\nu_2 - (\lambda_2 + p_1\nu_1 r_{12})y)^2}(p_2 P(1,0) - p_1 P(0,y)) \\
&+ \frac{p_1\nu_2\lambda_2 + (2p_1 - 1)\nu_1\nu_2 r_{12}}{(p_2\nu_2 - (\lambda_2 + p_1\nu_1 r_{12})y)^2}P(0,0) \\
&- \frac{p_1(\nu_2 + \nu_1 r_{12}y)}{p_2\nu_2 - (\lambda_2 + p_1\nu_1 r_{12})y}\frac{\mathrm{d}}{\mathrm{d}y}P(0,y).
\end{aligned}
\tag{10.9}
$$

Using

$$
p_2 P(1,0) - p_1 P(0,1) = \frac{p_2\gamma_1}{\nu_1} - \frac{p_1\gamma_2}{\nu_2} + (1 - 2p_1)P(0,0),
\tag{10.10}
$$

we set $y = 1$ in (10.9) and obtain after some rewriting (for $p_2\nu_2 \neq \lambda_2 + p_1\nu_1 r_{12}$)

$$
\begin{aligned}
\mathbb{E}(X_2) &= \left[\frac{\mathrm{d}}{\mathrm{d}y}P(1,y)\right]_{y=1} \\
&= \frac{\gamma_2}{p_2\nu_2 - (\lambda_2 + p_1\nu_1 r_{12})} \\
&- \frac{p_1(\nu_2 + \nu_1 r_{12})}{p_2\nu_2 - (\lambda_2 + p_1\nu_1 r_{12})}\left[\frac{\mathrm{d}}{\mathrm{d}y}P(0,y)\right]_{y=1}.
\end{aligned}
\tag{10.11}
$$

Thus, to determine $\mathbb{E}X_1$ and $\mathbb{E}X_2$, we need to compute $[\frac{\mathrm{d}}{\mathrm{d}y}P(0,y)]_{y=1}$. Note that for $p_1 = p$, $r_{12} = 1$, $r_{21} = 0$ and $\lambda_2 = 0$, (10.11) reduces to (9.38).

The case that $p_2\nu_2 = \lambda_2 + p_1\nu_2 r_{12}$ is special. Substituting $y = 1$ into (10.8) yields after applying l'Hôpital's rule:

$$
1 = \frac{\gamma_1 r_{12} + \lambda_2 - p_2\nu_2 - (\nu_2 - \lambda_2)\left[\frac{\mathrm{d}}{\mathrm{d}y}P(0,y)\right]_{y=1}}{-\lambda_2 - p_1\nu_1 r_{12}}.
\tag{10.12}
$$

Since $\gamma_1 r_{12} + \lambda_2 = \gamma_2$ this gives after some rewriting

$$\left[\frac{d}{dy}P(0,y)\right]_{y=1} = \frac{\gamma_2}{\nu_2 - \lambda}. \tag{10.13}$$

10.4 Preemptive Priority

When the proportion of the service capacity p_1 is set to one, the jobs at station 1 have preemptive priority over the jobs at station 2. For this case the generating function $P(x,y)$ can be obtained without employing the theory of boundary value problems, as will be shown in this section.

The functional equation (10.4) then reduces to

$$h_1(x,y)P(x,y) = h_3(x,y)P(0,y) + h_4(x,y)P(0,0). \tag{10.14}$$

Let $x = \xi(y)$ denote the unique solution of $h_1(x,y) = 0$ within the unit circle. That is

$$\xi(y) = \frac{\nu_1(1 - r_{12} + r_{12}y)}{\lambda_1(1 - \xi(y)) + \nu_1 + \lambda_2(1 - y)}. \tag{10.15}$$

For $x = \xi(y)$, the right-hand side of (10.14) should equal zero, yielding

$$P(0,y) = \frac{\nu_2(r_{21}\xi(y) - y + 1 - r_{21})P(0,0)}{y(\lambda_1\xi(y) - \lambda - \nu_2) + \nu_2(r_{21}\xi(y) + 1 - r_{21}) + \lambda_2 y^2}. \tag{10.16}$$

Since $P(0,0) = 1 - \rho_1 - \rho_2$, substituting (10.16) into (10.14) gives an expression for $P(x,y)$ in terms of $\xi(y)$ only, i.e.

$$P(x,y) = \frac{1}{h_1(x,y)}\left[\frac{h_3(x,y)\nu_2(r_{21}\xi(y) - y + 1 - r_{21})P(0,0)}{y(\lambda_1\xi(y) - \lambda - \nu_2) + \nu_2(r_{21}\xi(y) + 1 - r_{21}) + \lambda_2 y^2}\right.$$
$$\left. + h_4(x,y)P(0,0)\right]. \tag{10.17}$$

We will now show that (10.15) and (10.16) have probabilistic interpretations.

Let N represent the number of jobs served during a busy period of station 1, and Y the number of external arrivals at station 2 (with rate λ_2) during a busy period of station 1. Then, whenever station 1 empties, the service at station 2 is restarted, and a certain number of jobs has arrived there. Denote this number by H, which may be represented as

$$H = Y + \sum_{i=1}^{N} Z_i, \tag{10.18}$$

where Y and N independent, and $Z_i = 1$ w.p. r_{12} and $Z_i = 0$ w.p. $1 - r_{12}$. With B_1 the service time of a job at station 1, the pgf of H is given by

$$
\begin{aligned}
H(y) &= \int_{t=0}^{\infty} \mathbb{E}(y^H|B_1 = t)\nu_1 e^{-\nu_1 t}dt \\
&= \int_{t=0}^{\infty} \sum_{n=0}^{\infty} \frac{(\lambda_1 t)^n}{n!} e^{-\lambda_1 t} \sum_{k=0}^{\infty} \frac{(\lambda_2 ty)^k}{k!} e^{-\lambda_2 t} \\
&\qquad \times (1 - r_{12} + r_{12}y) \cdot \mathbb{E}(y^{H_1 + \cdots + H_n})\nu_1 e^{-\nu_1 t}dt \\
&= (1 - r_{12} + r_{12}y)\nu_1 \int_{t=0}^{\infty} e^{t(\lambda_1 H(y) - \lambda + \lambda_2 y - \nu_1)}dt,
\end{aligned}
$$

which matches (10.15).

To explain (10.16), we introduce the random vector (Y_1, Y_2), where Y_1 and Y_2 represent the stationary number of jobs at stations 1 and 2, respectively, at a point in time during a busy period of station 1. With $Q(x, y)$ the joint pgf of (Y_1, Y_2), it can be shown that

$$
Q(x, y) = \frac{x(\nu_1 - \lambda_1)(x - H(y))}{(\lambda + \nu_1)x - \lambda_1 x^2 - (1 - r_{12})\nu_1 - r_{12}\nu_1 y - \lambda_2 xy}. \tag{10.19}
$$

With $X_2^{(i)}$ the stationary number of jobs at station 2 during idle periods of station 1, we have that

$$
(X_1, X_2) = \begin{cases}
(0, X_2^{(i)}), & \text{w.p. } q_1, \\
(0, X_2^{(i)}) + (Y_1, Y_2), & \text{w.p. } q_2, \\
(0, X_2^{(i)} - 1|X_2^{(i)} > 0) + (Y_1, Y_2), & \text{w.p. } q_3,
\end{cases} \tag{10.20}
$$

where q_1, q_2, q_3 denote the probabilities that an arbitrary time point falls within an idle period of station 1, within a busy period of station 1 that is started with an external arrival to station 1, and within a busy period of station 1 that is started with a job coming from station 2, respectively. That is,

$$
q_1 = 1 - \rho_1, \quad q_2 = \rho_1 \frac{\lambda_1(1 - \rho_1)}{\lambda_1(1 - \rho_1) + \nu_2 r_{21}\rho_2}, \quad q_3 = \rho_1 \frac{\nu_2 r_{21}\rho_2}{\lambda_1(1 - \rho_1) + \nu_2 r_{21}\rho_2}.
$$

From (10.20) we see that

$$
P(x, y) = q_1 \frac{P(0, y)}{P(0, 1)} + q_2 \frac{P(0, y)}{P(0, 1)} Q(x, y) + q_3 \frac{1}{y}\left(\frac{P(0, y) - P(0, 0)}{P(0, 1) - P(0, 0)}\right) Q(x, y), \tag{10.21}
$$

which, after some lengthy calculations, can be shown to be equal to (10.16).

The above derivation of $P(x, y)$ can be extended to generally distributed service times, since in that case the decomposition in (10.20) continues to hold.

10.5   Boundary Value Problem

From now on we assume that $p_1 \neq 0$, $p_2 \neq 0$. In the analysis of the functional equation (10.4) a crucial role is played by the kernel $h_1(x, y)$. Due to the regularity properties of $P(x, y)$, for each pair (x, y) on and within the unit circle for which $h_1(x, y)$ equals zero, the right-hand side of (10.4) must vanish. This provides us with a relation between the unknown functions $P(0, y)$ and $P(x, 0)$. Blanc [18] has studied the transient behaviour of the two-station network without coupled processors. For this model, the kernel $h_1(x, y)$ is of the exact same form, and most of the results presented in this section also follow from his work.

Observe that $h_1(x, y)$ is for each x a polynomial of degree 2 in y, i.e.

$$h_1(x, y) = a(x)y^2 + b(x)y + c(x), \tag{10.22}$$

where

$$
\begin{aligned}
a(x) &= -\lambda_2 x - p_1\nu_1 r_{12}, \\
b(x) &= (\lambda + p_1\nu_1 + p_2\nu_2)x - \lambda_1 x^2 - p_1\nu_1(1 - r_{12}), \\
c(x) &= -p_2\nu_2 r_{21}x^2 - p_2\nu_2(1 - r_{21})x.
\end{aligned}
$$

For every x, there are two possible values of y, $y_1(x)$ and $y_2(x)$ say, such that $h_1(x, y_1(x)) = h_1(x, y_2(x)) = 0$. These values can be described by the two-valued function

$$y(x) = \frac{1}{2a(x)}(-b(x) \pm \sqrt{D(x)}), \tag{10.23}$$

where

$$D(x) = b^2(x) - 4a(x)c(x).$$

LEMMA 10.2 *The algebraic function $y(x)$, defined by $h_1(x, y(x)) = 0$, has four real branch points $0 \leq x_1 < x_2 \leq 1 < x_3 < x_4$.*

Proof The branch points of $y(x)$ are zeros of the discriminant $D(x)$. We have that $D(0) = (p_1\nu_1(1 - r_{12}))^2$ and $D(1) = (\lambda_2 + p_1\nu_1 r_{12} - p_2\nu_2)^2$. Assuming $D(0)$, $D(1)$ and λ_1 larger than zero, we have that $D(x) > 0$ as $x \downarrow 0$, $x = 1$ and $x \to \infty$. On the other hand, the function $b(x)$ is negative as $x \downarrow 0$ and $x \to \infty$, but positive at $x = 1$. Hence, $b(x)$ has one zero in the interval $(0, 1)$, and one zero in the interval $(1, \infty)$. At these points, $D(x) = -4a(x)c(x) < 0$. The remaining cases are left to the reader. $\square$

For later use, we now study the mapping $y(x)$ for $x \in [x_1, x_2]$ in some more detail. This mapping can be shown to give rise to a smooth and closed contour L, as specified in the next lemma.

LEMMA 10.3 *For each $x \in [x_1, x_2]$, $y(x)$ lies on the closed contour L, which is symmetric with respect to the real line. For $p_{12} = p_{21} = 0$, L is defined by $|y(x)|^2 = p_2\nu_2/\lambda_2$. Otherwise, $|y(x)|^2$ can be written as a function of $\mathrm{Re}(y)$, and*

$$|y|^2 \leq \frac{c(x_2)}{a(x_2)}. \tag{10.24}$$

Proof For $x \in [x_1, x_2]$, $D(x)$ is negative, so $y_1(x)$ and $y_2(x)$ are complex conjugates. It also follows that $|y(x)|^2 = c(x)/a(x)$, which together with

$$\frac{\mathrm{d}}{\mathrm{d}x}\left[\frac{c(x)}{a(x)}\right] = \frac{p_2\nu_1 r_{12}(1 - r_{21} + 2r_{21}x + r_{21}\lambda_2 x^2)}{p_1\nu_1 r_{12} + \lambda_2 x} \tag{10.25}$$

being nonnegative for $x \in (0, \infty)$ proves (10.24).

We can further solve $|y(x)|^2 = c(x)/a(x)$ as a function of x, and denote the solution that lies within $[x_1, x_2]$ by $\tilde{x}(y)$, i.e.

$$\tilde{x}(y) \ := \ \frac{\lambda_2|y|^2 - p_1\nu_1(1 - r_{21})}{2p_2\nu_2 r_{21}} \tag{10.26}$$
$$\frac{-\sqrt{(p_1\nu_1(1 - r_{21}) - \lambda_2|y|^2)^2 + 4p_1\nu_1 r_{12}p_2\nu_2 r_{21}|y|^2}}{2p_2\nu_2 r_{21}}.$$

So $\tilde{x}(y)$ is in fact the one-valued inverse function of $y(x)$. For each $y \in L$ it then holds that

$$\mathrm{Re}(y) = \frac{-b(\tilde{x}(y))}{2a(\tilde{x}(y))}. \tag{10.27}$$

Solving (10.27) as a function of $|y(x)|^2$ then gives an expression for $|y(x)|^2$ in terms of $\mathrm{Re}(y)$. □

We will henceforth denote the interior of L by L^+, and set $\alpha := y(x_2) = c(x_2)/a(x_2)$, representing the point on L with the largest modulus.

We will now show how the zero-set leads to a Riemann–Hilbert problem for the function $P(0, y)$.

LEMMA 10.4 *The function $P(0, y)$ is regular in the domain L^+ and satisfies for $y \in L$ the condition*

$$\mathrm{Im}(P(0, y)) = \mathrm{Im}\left(-P(0, 0)\frac{h_4(\tilde{x}(y), y)}{h_3(\tilde{x}(y), y)}\right). \tag{10.28}$$

Proof For zero-pairs (x, y) of the kernel for which $P(x, y)$ is finite we have

$$h_2(x, y)P(x, 0) + h_3(x, y)P(0, y) + h_4(x, y)P(0, 0) = 0, \tag{10.29}$$

from which it follows that

$$P(0, y) = \frac{1-p}{p} P(x, 0) - \frac{h_4(x, y)}{h_3(x, y)} P(0, 0). \tag{10.30}$$

Thus, (10.28) follows from the fact that $P(x, 0)$ is real for $x \in [x_1, x_2]$. If $\alpha \leq 1$, L lies entirely within the unit circle. Hence, $P(0, y)$ is regular in L^+. If $\alpha > 1$, $P(0, y(x))$ can be continued analytically over the interval $[x_1, x_2]$ via (10.29), because $P(x, 0)$ is regular on this interval. Hence, the analytic continuation of $P(0, y)$ is finite at $y = y(x_2)$. Because $P(0, y)$ has a power series expansion at $y = 0$ with positive coefficients, this implies that $P(0, y)$ is regular for $|y| < y(x_2)$ and hence in L^+. $\qquad \square$

Lemma 10.4 shows that determining $P(0, y)$ reduces to the following Riemann–Hilbert boundary value problem on the contour L: Determine a function $P(0, y)$ such that

1. $P(0, y)$ is regular for $y \in L^+$ and continuous for $y \in L^+ \cup L$.

2. $\mathrm{Re}\, (iP(0, y)) = \chi(y)$, for $y \in L$,

where

$$\chi(y) = \mathrm{Im} \left(P(0, 0) \frac{h_4(x(y), y)}{h_3(x(y), y)} \right).$$

As done before, we transform the boundary condition (10.28) to a condition on the unit circle, see, e.g. Muskhelishvili [129], p. 108. Denote the unit circle by C and its interior by C^+. We introduce the conformal mapping:

$$z = f(y) : L^+ \rightarrow C^+, \tag{10.31}$$

and its inverse

$$y = f_0(z) : C^+ \rightarrow L^+. \tag{10.32}$$

Using these mappings, we can reduce the Riemann–Hilbert problem on L to the following problem: Determine a function $G(z)$ such that

1. $G(z)$ is regular for $z \in C^+$ and continuous for $z \in C \cup C^+$.

2. $\mathrm{Re}\, (iG(z)) = \bar{\chi}(z)$, for $z \in C$, where $\bar{\chi}(z) = \chi(f_0(z))$.

The above problem is known as the Dirichlet problem on a circle. Its solution is given by, see Muskhelishvili [129], p. 108,

$$G(z) = -\frac{1}{2\pi} \oint_C \bar{\chi}(w) \frac{w + z}{w - z} \frac{dw}{w} + K_1, \quad z \in C \cup C^+, \tag{10.33}$$

with K_1 some real constant. In this way, $P(0, y) = G(f(y))$ has been formally determined as

$$P(0, y) = -\frac{1}{2\pi} \oint_C \bar{\chi}(w) \frac{w + f(y)}{w - f(y)} \frac{dw}{w} + K_1, \quad y \in L \cup L^+. \quad (10.34)$$

In the general case, in order to evaluate $\bar{\chi}(w)$, the mapping $f_0(z)$ should be determined using the procedure as described in Sect. 9.5. The procedure consists of finding a fixed number of boundary correspondence points by numerically solving Theodorsen's integral equation. Exceptions are the case that $r_{12} = 1, r_{21} = \lambda_2 = 0$, see Remark 9.1, and the case of two parallel queues discussed below.

10.5.1   The Case $r_{12} = r_{21} = 0$

In case $r_{12} = r_{21} = 0$, there is no routing of customers between the stations, and the network is reduced to two parallel queues. As mentioned before, this model has been analysed in Fayolle and Iasnogorodski [61]. Since the contour L is in this case a circle, $|y(x)|^2 = p_2 \nu_2 / \lambda_2$, the mappings $f(z)$ and $f_0(z)$ are simply given by

$$f(y) = \frac{y}{\sqrt{p_2 \nu_2 / \lambda_2}}, \quad f_0(y) = y\sqrt{p_2 \nu_2 / \lambda_2}. \quad (10.35)$$

It then readily follows that, for $y \in L \cup L^+$,

$$\begin{aligned}
P(0, y) &= -\frac{1}{2\pi} \oint_C \chi(w\sqrt{p_2\nu_2/\lambda_2}) \frac{w + f(y)}{w - f(y)} \frac{dw}{w} + K_1 \\
&= -\frac{i}{2\pi} \int_{-\pi}^{\pi} \chi(e^{i\phi}\sqrt{p_2\nu_2/\lambda_2}) \frac{e^{i\phi} + f(y)}{e^{i\phi} - f(y)} d\phi + K_1.
\end{aligned}$$
$$\qquad\qquad (10.36)$$

The case $r_{12} = r_{21} = 0$ is the only situation for which we cannot specify the inverse function $\tilde{x}(y)$ according to (10.26). Instead, we can use the fact that L is a circle to derive $\tilde{x}(y)$. We start from the observation that for the zero-pairs (x, y) of the kernel $h_1(x, y)$ it holds

$$(\lambda + p_1\nu_1 + p_2\nu_2)x - \lambda_1 x^2 - \lambda_2 xy - p_1\nu_1 - p_2\nu_2\frac{x}{y} = 0. \quad (10.37)$$

Also, for each $y \in L$ we have

$$y = \sqrt{\frac{p_2\nu_2}{\lambda_2}} e^{i\phi} = \sqrt{\frac{p_2\nu_2}{\lambda_2}}(\cos\phi + i\sin\phi), \quad (10.38)$$

for some $\phi \in [0, 2\pi]$. Plugging (10.38) into (10.37) and solving for x gives two possible outcomes

$$x_1(\phi) = \frac{\gamma - \sqrt{\gamma^2 - 4p_1\nu_1\lambda_1}}{2\lambda_1}, \quad x_2(\phi) = \frac{\gamma + \sqrt{\gamma^2 - 4p_1\nu_1\lambda_1}}{2\lambda_1}, \quad (10.39)$$

where $\gamma := \lambda + p_1\nu_1 + p_2\nu_2 - 2\sqrt{p_2\nu_2\lambda_2}\cos\phi$. It is straightforward to see that $x_1(\phi) \in [x_1, x_2]$ for all $\phi \in [0, 2\pi]$. Therefore, the inverse function $\tilde{x}(y)$ is for all $y \in L$ specified by

$$\tilde{x}(e^{i\phi}\sqrt{p_2\nu_2/\lambda_2}) = x_1(\phi). \quad (10.40)$$

10.6 Conclusion and Further Research

For the two-station open queueing network with Poisson arrivals, exponential service times, and coupled processors, we have shown that the pgf of the joint stationary queue length distribution can be found using the theory of boundary value problems.

Calculating the performance measures described in Sect. 10.3 involves computational issues as described in Sect. 9.6 for the tandem queue. As for the tandem queue, it is crucial to determine either the mapping $f(z)$ or its inverse mapping $f_0(z)$. If no explicit description of either mapping is available, the inverse mapping $f_0(z)$ can be determined numerically using the procedure described in Sect. 9.5. So far, we have not been able to derive an explicit description for $f(z)$ or $f_0(z)$, leaving it as a challenging topic for further research.

Next to coupled processors, we introduced in Sect. 2.4 a scheduling discipline referred to as *partial coupling*, which, in the more general context of this chapter, would imply the following. Whenever both queues are nonempty, the capacity is divided among stations 1 and 2 according to fixed fractions. Whenever queue 1 is empty, all capacity goes to station 2. When queue 2 is empty, however, the capacity of station 1 is not increased. Partial coupling is an example of a non-work-conserving scheduling discipline. More generally, we could describe the class of non-work-conserving scheduling disciplines as follows. When both stations are nonempty, the service rate of station j is equal to ν_j. If one of the stations becomes empty, the service rate at the other station changes from ν_j to ν_j^*, where $\nu_j^* \neq \nu_1 + \nu_2$ for some $j = 1, 2$.

For the two-station open queueing network with Poisson arrivals, exponential service times, and a non-work-conserving scheduling discipline, again a Riemann–Hilbert boundary value problem can be formulated on some contour M (with M^+ the interior of M): Determine the function $P(0, y)$ such that

1. $P(0, y)$ is regular for $y \in M^+$ and continuous for $y \in M^+ \cup M$.

2. $\mathrm{Re}\,(p(y)P(0, y)) = q(y)$, for $y \in M$.

Note that due to the function $p(y)$, this is a more general Riemann–Hilbert boundary value problem than those we have encountered earlier, in which case $p(y) = 1$. Solving this more general Riemann–Hilbert boundary value problem requires a slightly more complicated technique, as explained in Mushkelisvili [129], p. 100, and Cohen and Boxma [39], p. 56. A detailed study of this more general Riemann–Hilbert boundary value problem is an interesting topic for further research.

PART V

EPILOGUE

Chapter 11

CABLE NETWORKS REVISITED

Reservation procedures consist of two phases: A request phase and a data-transmission phase. In the previous chapters, we have proposed models for the delay in the request phase, see Chap. 4, and for the delay in the data-transmission phase, see Chaps. 7 and 8. Along the way, we have made some suggestions to improve the standard algorithms to schedule the reservation procedure.

In this chapter, we tie our insights together by combining the models to derive an approximation of the total average packet delay in a reservation procedure. We also combine the proposed scheduling improvements into one 'enhanced' scheduling strategy. Then we present quantitative results for two different objectives. Firstly, we compare the approximation of the total average packet delay with delay figures obtained by system simulations for cable networks. Thus, we assess the extent to which our approximation has numerical significance for cable networks. Secondly, we compare the delay when using the enhanced reservation procedure to the delay when using a more standard scheduling. This enables us to quantify the effect of the scheduling improvements suggested in this monograph.

11.1   Introduction

In the previous chapters, we have proposed models that are useful in the analysis of a reservation procedure, where the requests are carried out in contention via a contention tree. In particular, we have used modifications of the repairman model to approximate the request delay in Chap. 4. The delayed bulk service queue was proposed in Chap. 8 as an appropriate model for the data queue of the reservation procedure.

Along with these models, we have made some suggestions to improve the standard algorithms for the reservation procedure. Specifically, in Sect. 4.2.1

we proposed scheduled access as a channel access protocol that is better than the more common access protocols of blocked and free access. In Sect. 8.1.1, a bandwidth allocation strategy was outlined that guarantees a certain minimum number of slots, c, per frame to the reservation process. In this strategy, c was determined on the basis of the total traffic intensity rather than fixed a priori.

These models were motivated by the use of reservation procedures in cable networks, and by the perceived deficiencies of the performance models that have been proposed for this application. At least qualitatively, we have been successful in this respect. Our model does capture a finite-population effect, via the approximation to the request delay (4.10) and incorporates a scheduling effect, see Figs. 8.6, 8.8, 8.10, and 8.12.

In this chapter, we tie the two models together, via some back-of-the-envelope calculations, and give an approximation to the total average packet delay in a reservation procedure. Moreover, we combine the proposed scheduling improvements into one scheduling strategy, which we will call the enhanced reservation procedure.

The chapter is organised as follows. In Sect. 11.2, we detail the traffic model, and in Sect. 11.3 we give the calculations that lead to an approximation to the total average packet delay. In Sect. 11.4, we carry out some numerical experiments that compare the proposed models for the mean delay with results obtained by simulation. Moreover, we quantify the effect of the improvements. These results suggest a number of directions for further research which we summarise in Sect. 11.5.

11.2   Traffic Model

We consider a very simple traffic model, that was also used in the simulations described in Sect. 1.4. In particular, we assume the following set-up:

- The population consists of N stations.

- The stations generate traffic according to independent Poisson processes with rate $\lambda = \Lambda/N$.

- The contention resolution is carried out by contention trees with one of the channel access protocols that were described in Sect. 4.2: Blocked access, free access, or scheduled access.

The total packet delay consists of two parts: The request delay and the data-transmission delay, corresponding to the two phases of the reservation procedure. To approximate the request delay, we use the results from Chap. 4. To approximate the data-transmission delay, we build upon Chap. 8. However, to use these latter results, we must specify the properties of the request size. The request size is the number of packets for which transmission is requested, and corresponds to the new arrivals, Y_n, in (8.4). In this section, we analyse the moments of the request size. These are then used, in Sect. 11.3, to calculate total packet delay.

In order to calculate these moments, we first consider the request size in a successful request, which we denote by R. From the Poisson assumption in the traffic model, it follows that

$$R \stackrel{d}{=} 1 + \text{Poisson}(\lambda S), \tag{11.1}$$

where S is the request delay. Thus R equals 1, for the packet that starts the contention process, plus the number of arrivals during the interval $[0, S]$. With $f_S(t)$ the density of the sojourn time, the joint density of (S, R) is given by

$$p(t, r) = e^{-\lambda t} \frac{(\lambda t)^{r-1}}{(r-1)!} f_S(t), \tag{11.2}$$

for $t \geq 0$ and $r \geq 1$. Taking expectations in (11.1), we obtain

$$\mathbb{E}(R) = 1 + \lambda \mathbb{E}(S), \tag{11.3}$$

and an approximation to $\mathbb{E}(S)$ is given by (4.10) in Chap. 4.

We now turn to the variance of R. Using P to denote a random variable distributed according to a $\text{Poisson}(\lambda S)$ distribution, we find that

$$\begin{aligned}
\mathbb{E}(R^2) &= \mathbb{E}((1+P)^2) \\
&= \mathbb{E}(1 + 2P + P^2) \\
&= 1 + 3\lambda \mathbb{E}(S) + \lambda^2 \mathbb{E}(S^2),
\end{aligned}$$

and consequently,

$$\text{var}(R) = \lambda \mathbb{E}(S) + \lambda^2 \text{var}(S). \tag{11.4}$$

In the ensuing approximation in the next section, we need to consider the request size in one time slot, rather than the request size in a successful request. For this, we must account for two further properties of the reservation process in cable networks that were both discussed in Sect. 1.3. Firstly, recall that one such time slot consists of three reservation minislots. Secondly, in each such minislot, there is, approximately, a probability ν of a successful request. Values for ν are given in Sect. 4.7. Ignoring the correlation between successive reservation minislots, it follows that approximately

$$Y \stackrel{d}{=} Y_1 + Y_2 + Y_3, \tag{11.5}$$

where the Y_j are independent and identically distributed, and where

$$Y_1 \stackrel{d}{=} B \cdot R. \tag{11.6}$$

In (11.6), B is a Bernoulli random variable, that equals 1 with probability ν and 0 with probability $1 - \nu$, and R is the request size in a successful request. Thus, with probability ν there is a successful request and the request size equals R. With probability $1 - \nu$, the request attempt fails, so that $Y_1 = 0$.

Above, we have calculated the moments of R and these can in turn be used to compute the moments of Y. A brief derivation shows that

$$\mathbb{E}(Y) = 3\nu(1 + \lambda\mathbb{E}(S)), \tag{11.7}$$

and

$$\text{var}(Y) = 3\nu\left(\lambda\mathbb{E}(S) + \lambda^2\text{var}(S)\right) + 3\nu(1 - \nu)\left(1 + \lambda\mathbb{E}(S)\right)^2. \tag{11.8}$$

11.3   Total Average Packet Delay

We propose to model the total packet delay, D_T as

$$D_T = 1 + d + D_R + D_Q. \tag{11.9}$$

Here, d is the minimum delay experienced by a packet due to the round-trip times in a cable network, D_R is the packet delay due to the request procedure, and D_Q is transmission delay due to the queueing at the data queue. That is, D_Q is the delay due to the queueing at the central scheduler, once the request has been successfully transmitted. As noted in Sect. 1.3, d equals approximately two frames in cable networks.

The packet delay, D_R, is the delay experienced by a random packet. As a random packet is more likely to belong to a larger request, the density of the sojourn time and request size of such a random packet is given by

$$\tilde{p}(t, r) = rp(t, r)/\mathbb{E}(R), \tag{11.10}$$

where $p(t, r)$ is given in (11.2), and where $t \geq 0$ and $r \geq 1$. In a successful request, one packet suffers the full request delay, whereas the remaining packets suffer a request delay that is uniformly distributed over S. Hence, using $(\tilde{S}, \tilde{R})$ to denote a pair of random variables with density (11.10):

$$D_R \stackrel{d}{=} \tilde{B} \cdot \tilde{S} + (1 - \tilde{B})U_{\tilde{S}},$$

where $U_{\tilde{S}}$ is a random variable that is uniformly distributed on $[0, \tilde{S}]$, and where $\tilde{B}$ is a binary random variable, such that:

$$\tilde{B} = \begin{cases} 1 & \text{w.p. } 1/\tilde{R}, \\ 0 & \text{w.p. } 1 - 1/\tilde{R}. \end{cases}$$

Consequently, we obtain

$$\mathbb{E}(D_R) = E\left(\frac{1}{\tilde{R}}\tilde{S} + \frac{\tilde{R}-1}{\tilde{R}}U_{\tilde{S}}\right)$$

$$= \frac{\mathbb{E}(S)}{1 + \lambda\mathbb{E}(S)} + \frac{\lambda\mathbb{E}(S^2)}{2(1 + \lambda\mathbb{E}(S))}. \tag{11.11}$$

We now turn to the transmission delay, D_Q. Recall that the delayed bulk service queue was used as a model for the data queue in a reservation procedure. This queue is served with a speed of s packets per frame, so that it takes approximately X/s frames to clear a data queue of size X. Consequently, we obtain that

$$\mathbb{E}(D_Q) \approx \mathbb{E}(X^d)/s, \tag{11.12}$$

where $\mathbb{E}(X^d)$ is the expected data-queue size in the delayed bulk service queue. Hence, an approximation to the expected transmission delay can be obtained from the approximation to the expected data-queue size given in (8.17). This approximation involves the moments of Y, which is the number of packets requested in one reservation slot. For these, we substitute the expressions given in (11.7) and (11.8).

By adding the approximations to the average packet request delay and the average data-transmission delay, we have expressed the total average packet delay in terms of the moments of the request delay S and the probability ν of a successful request. As argued extensively in Chap. 4, the moments of the request delay depend on the access method that complements the basic contention tree: Free access, blocked access, or scheduled access. Moreover, it was shown in Sect. 4.7 that the success probability depends on the access method as well, and that free access and scheduled access have slightly higher success probability than blocked access.

Thus, we obtain an approximation for the total average packet delay as a function of the traffic intensity. In addition, the approximation makes it clear that the total average packet delay depends on the channel access protocols in two respects: Through the service rate ν and through the variance of the request delay.

Before we turn to the numerical comparisons in Sect. 11.4, three comments are in order.

REMARK 11.1 In order to actually calculate the moments of the sojourn times in the machine repair models, we must specify the service rate with which the repairman operates. Here, we argue as follows. There are on average $f - \Lambda$ slots per frame that are used for the contention process. It follows that the appropriate frame rate is

$$\mu := 3\nu(f - \Lambda),$$

where ν is the rate of the contention resolution process per minislot.

REMARK 11.2 In Sect. 11.4 below, we make a number of approximations. These are all based on the explicit expressions that are obtained when considering the contention process in heavy traffic. For these approximations to be applicable, it is necessary that

$$\mu < \Lambda,$$

where Λ is the traffic intensity per frame. Substitution of μ then shows that we must require that

$$\Lambda > \frac{3\nu}{1 + 3\nu} f. \tag{11.13}$$

Using the values for ν as given in, e.g. Sect. 4.7 then shows that our approximations are valid for traffic loads above approximately 55%. Still, the performance curves show the full range of traffic intensities. These are obtained by thresholding the approximations to zero, if condition (11.13) fails.

REMARK 11.3 In deriving the expected transmission delay, we have ignored the detailed arrival and departure pattern of packets within a frame; see Chap. 7 for more detailed calculations in case $d = 0$.

11.4   Numerical Assessment

In this section, we carry out a numerical assessment of the models and schedules considered in this monograph. In Sect. 11.4.1, we investigate the quality of our approximation to the total average packet delay by means of a comparison with results obtained by simulations of cable networks. This is done for blocked contention trees and the three network scenarios introduced in Sect. 1.4.

In Sect. 11.4.2, we assess the gain from using the schedules in this monograph. In particular, we calculate the expected delay when using the standard blocked and free contention trees with a fixed minimal amount guaranteed to the contention process. We benchmark the performance of the best schedule developed in this monograph against these standard schedules. In this best schedule, we have used the scheduled access protocol and have chosen c (the minimum number of slots per frame for the reservation process) optimally depending on the traffic intensity.

11.4.1   Approximation to the Total Average Packet Delay

In an experiment discussed in Sect. 1.4.1 we compared the performance of cable networks in three different network scenarios. In these scenarios, we varied the number of stations in the network. Performance curves for this experiment were presented in Fig. 1.2 for the particular case that there are 50, 100, and 200

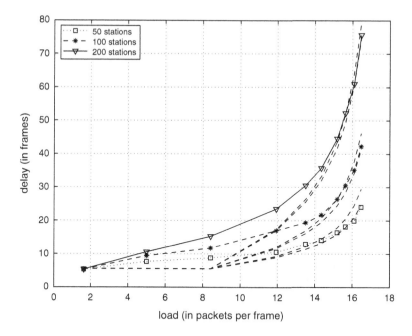

Fig. 11.1. Total average packet delay obtained by simulation (*solid lines*) and approximating upper and lower bounds (*dashed lines*) for three network scenarios with 50, 100, and 200 stations

stations in the network. These curves are displayed again in Fig. 11.1 as the solid lines. Additionally, Fig. 11.1 shows the approximating upper and lower bounds based on the approximation described in Sect. 11.3. These approximations are shown as dashed lines.

Comparing the simulation results with the approximations, we see that there is excellent numerical agreement between theory and experiment for high traffic intensities, decreasing agreement for medium traffic intensities, and a complete mismatch for low traffic intensities.

These observations should cause no surprise. Our analyses that have led to the approximating upper and lower bounds are largely based on asymptotic arguments: Both the approximation to the variance of the request delay and the approximation to the expected data-queue size in a delayed bulk service queue are based on 'heavy traffic' arguments. These are valid for high traffic intensities only and one cannot expect the approximation to hold beyond this regime. Furthermore, our model is not sufficiently detailed to capture the system performance for light and medium traffic intensities. This is due to the presence of substantial round-trip times in combination with the use of the contention trees, as can be concluded from the following argument. The round-trip times hamper the expansion of the contention tree, as a conflict in frame i can only be resolved in frame $i + 3$. This causes the request delays to be much larger than predicted: Our model assumes that all free space can be used for the expansion of the current tree. This is a good approximation in heavy traffic, when the trees are large and there are only few slots per frame that can be devoted to contention resolution. However, this approximation fails in light and medium traffic, when the trees are small and the reservation slots are plentiful. We get back to this issue in the concluding section of this chapter.

11.4.2   The Enhanced Reservation Procedure

We compare our enhanced procedure to the two standard methods to organise the reservation procedure. There are two elements to our improvement. Firstly, we used the scheduled access protocol rather than the free or blocked access protocols. This scheduled access protocol improves on the blocked access protocol, as it operates at a higher rate. Moreover, the scheduled access protocol improves on both the free and the blocked access protocols, as the variance of the request delay is much smaller. Secondly, the enhanced reservation procedure schedules the amount of bandwidth allocated to the reservation process. In Sect. 8.1.1 we have introduced two such schedules: A static and an adaptive schedule. It was found in Chap. 8 that adaptive scheduling outperforms static scheduling, see Sect. 8.4.2. However, for static scheduling approximate, analytical, performance figures could be derived, whereas no such approximations are available for adaptive scheduling. Therefore, we confine attention to this static schedule, which uses a minimum amount of bandwidth guaranteed to the reservation process that is tuned to the traffic intensity. The number of slots guaranteed to the reservation is an integral number of slots per frame and is determined by minimising the expression in (8.17) given the traffic intensity. Here, we assume that this traffic intensity can be estimated reliably, so that this scheduling policy can indeed be implemented. The standard methods both guarantee one time slot to the reservation process, independent of the traffic conditions.

In Fig. 11.2, we show the performance curves for the three procedures. All curves are based on the approximations given in Sect. 11.3. Note that we only show the performance curves for medium and high traffic intensities, where our approximations are reasonably accurate. In comparing the three performance curves, we see that the enhanced algorithm is superior over this range of traffic intensities. For high traffic intensities, it reduces the total average packet delay by about 20% as compared to its best competitor, the blocked contention tree. The performance with the free contention trees is initially a little better than with the blocked contention trees. This is due to the higher efficiency of the free contention trees. At higher traffic intensities, however, the performance with free contention trees is much worse, due to large variances.

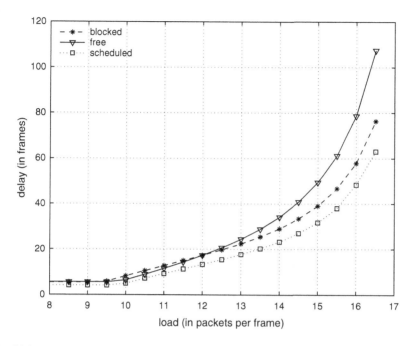

Fig. 11.2. Average total packet delay for standard organisation of reservation procedure with blocked trees or free trees with $c = 1$, and the enhanced 'scheduled' algorithm

11.5 Further Research

In this monograph we have proposed delay models for multiaccess, in case the multiaccess is organised via a reservation procedure and the requests are carried out in contention using contention trees. The models were motivated by multiaccess in cable networks. However, the formulation and analysis of the models have been given in mathematical terms, by appropriate modifications of some key performance models.

Because of this approach, the value of our contributions is not limited to cable networks. The proposed models are of sufficient generality to be useful in other contexts as well. However, we have also shown that this approach has led to results that have clear significance in the context of cable networks, and we have made a number of suggestions for improvement of the standard way to organise the reservation procedure. Here, we refer in particular to the modification of the tree procedure described in Sect. 4.2.1 and the schedules to determine the amount of bandwidth to be used for reservation described in Sect. 8.1.1. Additionally, the results from this chapter demonstrate that the results are numerically accurate. The approximations to the total average packet delay match the simulation values quite well in case of heavy traffic. Thus, we were also able to quantify the gain in implementing our suggestions, which we estimated to be an improvement of about 20% over the best standard competitor.

This monograph has also left a number of issues unresolved which we propose for future research. Some detailed suggestions concerning the analysis of the key models have been formulated along the way in the concluding sections of the various chapters. We now formulate a number of research questions that relate more specifically to the analysis of the reservation procedure.

It has been pointed out in Sect. 11.4.1 that the proposed approximations are valid only in the heavy traffic regime, i.e. from approximately 25% below the stability boundary up to the stability boundary. It is of interest to extend the approximations to encompass the other traffic regimes. The major obstacle towards this goal is the limitation of the proposed model for the request delay, as it does not account for the presence of propagation delay in combination with the use of the contention trees.

We see two possible ways to extend our models for trees to encompass the round-trip delay. Following a suggestion of Massey [119], it is possible to view the delayed communication channel as a number of interleaved non-delayed communication channels. Thus, we can confine the analysis to a non-delayed channel, which is easier to carry out. Of course, an implementation which splits the channel into a number of interleaved channels is less efficient than an implementation which does not split the channel, so that this approach is not fully satisfactory from an engineering perspective. A preferable approach then is to extend the system with an alternative access mode, such as direct access. In this case, one can use those time slots for direct access, which are neither guaranteed for reservation nor usable for data transmission of packets in the data queue. Thus, one will mitigate the problems with the model for the request delay. Moreover, there is numerical evidence, see Pronk et al. [145] and Hekstra-Nowacka et al. [78], that this mixing of access modes, for medium traffic intensities, is more efficient than the exclusive use of the reservation procedure.

In view of this latter evidence, it is natural to conjecture that direct access is best for light traffic, mixed access is best for medium traffic intensity, and that the reservation procedure is the most appropriate schedule in case of heavy traffic. From a system's viewpoint, we can thus conclude that the analysis of the reservation procedure in the heavy traffic case, as undertaken in this monograph, is indeed the most relevant one.

It is important to evaluate our scheduler enhancements in an actual system. Such an evaluation will certainly use traffic conditions that are different from the simplifying Poisson assumption that we have entertained throughout this monograph. Note, though, that the assumption of independent Poisson processes at the various stations constitutes the core of one of the two standard performance models that are used in the standardisation of cable access systems, see, e.g. [73, 74, 145, 150]. Moreover, the extension of our results to the second standard performance model, which assumes independent batch Poisson processes, seems not very difficult. However, the real challenge here is the incorporation of the burstiness which characterises real traffic. For an experimental approach to assess the effects of more realistic traffic models in cable networks, see Ivanovich et al. [83, 84].

Furthermore, in Chap. 8, we have indicated that the analysis of the delayed bulk service queue forms the basis of two scheduling strategies. In the enhanced algorithm, as evaluated in Sect. 11.4.2, we have confined ourselves to the static scheduler. However, a comparison of a reservation procedure based on the adaptive scheduling strategy is also called for, as is a comparison with the more specific, ad hoc, algorithms proposed in the context of cable networks, see, e.g. [73, 150].

To introduce a final topic for future research, note that in this chapter, we have approached the analysis of the reservation procedure by means of a decomposition of the problem into a request phase and a data-transmission phase. Separate models for both phases were proposed and analysed: The repairman model for the request phase and the delayed bulk service queue for the data-transmission phase. By combining the results, we have been able to approximate the average packet delay for the reservation procedure.

This approach is natural, but has its limitations as it does not generalise to higher moments or the distribution of the packet delay: In order to establish further properties of the packet delay, it is necessary to consider the detailed interaction between the two phases, caused by the correlation between request size and request delay.

So, a final topic for future research is the extension of our result on average packet delay to higher moments and distributions. As the foregoing argument shows, this will require a completely different approach from the decompositional one taken in this chapter. One possible path towards a more comprehensive understanding of reservation procedures is initiated in Part IV of this book, and utilises tandem queues. Another noteworthy attempt to formulate and analyse a more integrated model can be found in one of the initial contributions to the Pelican project (the joint Philips-EURANDOM project on conditional access in networks that also provided the impetus for our work): In Palmowski et al. [135, 136] an open, gated, tandem system is considered in which the service time in the second server of the tandem is coupled to the (batch) sojourn time in the first server of the tandem. This nicely captures the interaction described above, as the batch sojourn time in the first server can be interpreted as request delay and the service time at the second queue as request size. As such, [135, 136] provide an interesting starting point to further pursue the study of reservation procedures. However, to make this contribution really convincing for cable networks, it should be modified to a closed system and the propagation delay should find a place. Part IV of this monograph, as well as the papers [135, 136] then illustrate the mathematical challenges that go with the extension of our result to higher moments and distributions.

References

[1] Abate, J. and W. Whitt (1992). Numerical inversion of probability generating functions. *Operations Research Letters* **12**: 245–251.

[2] Abolnikov, L. and A. Dukhovny (1987). Necessary and sufficient conditions for the ergodicity of Markov chains with transition $\Delta_{m,n}$ ($\Delta'_{m,n}$)-matrix. *Journal of Applied Mathematics and Simulation* **1**: 13–24.

[3] Ackroyd, M. (1980). Computing the waiting time distribution for the G/G/1 queue by signal processing methods. *IEEE Transactions on Communications* **28**: 52–58.

[4] Adan, I.J.B.F., J.S.H. van Leeuwaarden, and E.M.M. Winands (2006). On the application of Rouché's theorem in queueing theory. *Operations Research Letters* **34**: 355–360.

[5] Adan, I.J.B.F. and Y.Q. Zhao (1996). Analyzing GI/E$_r$/1 queues. *Operations Research Letters* **19**: 183–190.

[6] Andradottir, S., H. Ayhan, and D. Down (2001). Server assignment policies for maximizing the steady-state throughput of finite queueing systems. *Management Science* **47**: 1421–1439.

[7] Asmussen, S. (2003). *Applied Probability and Queues* (2nd edition), Springer, New York.

[8] Bagchi, T.P. and J.G.C. Templeton (1970). *Numerical Methods in Markov Chains and Bulk Queues*, Springer, New York.

[9] Bailey, N.T.J. (1954). On queueing processes with bulk service. *Journal of the Royal Statistical Society, Series B* **16**: 80–87.

[10] Baskett, F., K.M. Chandy, R.R. Muntz, and F.G. Palacios (1975). Open, closed, and mixed networks of queues with different classes of customers. *Journal of the ACM* **22**: 248–260.

[11] Bayer, N. (1996). On the identification of Wiener-Hopf factors. *Queueing Systems* **23**: 293–300.

[12] Bellman, R. (1970). *Introduction to Matrix Analysis* (2nd edition), McGraw-Hill, London.

[13] Benini, L. and G. de Micheli (2002). Networks on Chip: A new SoC paradigm. *IEEE Computer* **35**: 70–78.

[14] Bertsekas, D.P. and R.G. Gallager (1992). *Data Networks*, Prentice-Hall, Englewood Cliffs, N.J.

[15] Bisdikian, C., K. Maruyama, D. Seidman, and D. Serpranos (1996). Cable access beyond the hype: On residential broadband data services over HFC networks. *IEEE Communications Magazine* **34**: 128–135.

[16] Blanc, J.P.C. (1982). *Application of the Theory of Boundary Value Problems in the Analysis of a Queueing Model with Paired Services*, Mathematical Centre Tract **153**, Amsterdam.

[17] Blanc, J.P.C. (1984). Asymptotic analysis of a queueing system with a two-dimensional state space. *Journal of Applied Probability* **21**: 870–886.

[18] Blanc, J.P.C. (1985). The relaxation time of two queueing systems in series. *Communications in Statistics-Stochastic Models* **1**: 1–16.

[19] Borst, S.C., O.J. Boxma, and M.J.G. van Uitert (2001). Two coupled queues with heterogeneous traffic. In: J.M. de Souza, N.L.S. da Fonseca, and E.A. de Souza e Silva (eds.) *Proceedings of ITC 17*, North-Holland, Amsterdam: 1003–1014.

[20] Boudreau, P.E., J.S. Griffin, and M. Kac (1962). An elementary queueing problem. *The American Mathematical Monthly* **69**: 713–724.

[21] Borst, S.C., O.J. Boxma, J.A. Morrison, and R. Núñez Queija (2003). The equivalence of processor sharing and service in random order. *Operations Research Letters* **31**: 254–262.

[22] Boxma, O.J., D. Denteneer, and J.A.C. Resing (2002). Some models for contention resolution in cable networks. In: E. Gregori, M. Conti, A.T. Campbell, G. Omidyar, and M. Zukerman (eds.) *Networking 2002*, Springer LNCS 2345, Berlin: 117–128.

[23] Boxma, O.J., D. Denteneer, and J.A.C. Resing (2003). Delay models for contention trees in closed populations. *Performance Evaluation* **53**: 169–185.

[24] Boxma, O.J. and W.P. Groenendijk (1988). Two queues with alternating service and switching times. In: O.J. Boxma and R. Syski (eds.) *Queueing Theory and its Applications, Liber Amicorum for J.W. Cohen*, North-Holland, Amsterdam: 261–282.

[25] Brockmeyer, E. (1948). A survey of A.K. Erlang's mathematical works. *Danish Academy of Technical Sciences* **2**: 101–126.

[26] Van den Broek, M.X. (2001). *On contention resolution procedures. Queuing analysis and simulation*, M.Sc. Thesis, Eindhoven University of Technology.

[27] Bruneel, H. (1984). A general model for the behaviour of infinite buffers with periodic service opportunities. *European Journal of Operational Research* **16**: 98–106.

[28] Bruneel, H. and B.G. Kim (1993). *Discrete-Time Models for Communication Systems including ATM*, Kluwer, Dordrecht.

[29] Capetanakis, J.I. (1977). *The Multiple Access Broadcast Channel: Protocol and Capacity Considerations*, Ph.D. dissertation, MIT, Dept. of Electrical Engineering and Computer Science, Cambridge, MA.

[30] Capetanakis, J.I. (1979). Tree algorithms for packet broadcast channels. *IEEE Transactions on Information Theory* **25**: 505–515.

[31] Chaudhry, M.L., C.M. Harris, and W.G. Marchal (1990). Robustness of rootfinding in single-server queueing models. *ORSA Journal on Computing* **3**: 273–286.

[32] Chaudhry, M.L. and J.G.C. Templeton (1983). *A First Course in Bulk Queues*, Wiley, New York.

[33] Coffman, E.G., G. Fayolle, and I. Mitrani (1986). Sojourn times in a tandem queue with overtaking: Reduction to a boundary value problem. *Communications in Statistics-Stochastic Models* **2**: 43–65.

[34] Cohen, J.W. (1975). The Wiener-Hopf technique in applied probability. In: J. Gani (ed.) *Perspectives in Probability and Statistics*, Applied Probability Trust, Sheffield: 145–156.

[35] Cohen, J.W. (1982). *The Single Server Queue*, 2nd edition, North-Holland, Amsterdam.

[36] Cohen, J.W. (1984). On a functional relation in three complex variables: Three coupled processors. *Technical Report Mathematical Institute Utrecht* 359, Utrecht University.

[37] Cohen, J.W. (1984). On processor sharing and random order of service (Letter to the editor). *Journal of Applied Probability* **21**: 937.

[38] Cohen, J.W. (1988). Boundary value problems in queueing theory. *Queueing Systems* **3**: 97–128.

[39] Cohen, J.W. and O.J. Boxma (1983). *Boundary Value Problems in Queueing System Analysis*, North-Holland, Amsterdam.

[40] Corner, M.D., J. Liebeherr, N. Golmie, C. Bisdikian, and D.H. Su (2000). A priority scheme for the IEEE 802.14 MAC protocol for hybrid fiber-coax networks. *IEEE/ACM Transactions on Networking* **8**: 200–211.

[41] Crommelin, C.D. (1932). Delay probability formulae when the holding times are constant. *Post Office Electrical Engineers Journal* **25**: 41–50.

[42] Crommelin, C.D. (1934). Delay probability formulae. *Post Office Electrical Engineers Journal* **26**: 266–274.

[43] Darroch, J.N. (1964). On the traffic light queue. *Annals of Mathematical Statistics* **35**: 380–388.

[44] Denteneer, D. (2001). A time sharing system with a waiting room. *National Laboratory Technical Note* TN-2001/168.

[45] Denteneer, D. (2001). Analysis of contention resolution in cable networks via an extension of the time sharing system. In: W. Verhaegh, J. Korst, and E. Aarts (eds.) *Scharm 2001, Philips Workshop on Scheduling and Resource Management*: 159–168.

[46] Denteneer, D. (2001). Efficient scheduling of contention trees. *National Laboratory Technical Note* TN-2001/169.

[47] Denteneer, D. (2005). *Data Transfer in Cable Networks: Delay Models for Multiaccess with Contention Trees*, Ph.D. thesis, Eindhoven University of Technology.

[48] Denteneer, D. (2006). Models for data transmission delay in cable networks. *Statistica Neerlandica* **60**: 12–45.

[49] Denteneer, D. and C. Gromoll (2007). Heavy traffic analysis of a closed queueing system with gated random order of service. In preparation.

[50] Denteneer, D., A.J.E.M. Janssen, and J.S.H. van Leeuwaarden (2005). Moments series inequalities for the discrete-time bulk service queue. *Mathematical Methods of Operations Research* **61**: 85–108.

[51] Denteneer, D., J.S.H. van Leeuwaarden, and J.A.C. Resing (2003). Bounds for a discrete-time multi-server queue with an application to cable networks. In: J. Charzinski, R. Lehnert, and P. Tran-Gia (eds.) *Providing Quality of Service in Heterogeneous Environments*, Proceedings of ITC **18**: 601–610.

[52] Denteneer, D. and J.S.H. van Leeuwaarden (2005). The delayed bulk service queue: A model for a reservation process. In: X. Liang, Z. Xin, V.B. Iversen, and G.S. Kuo (eds.) *Performance challenges for efficient next generation networks*, Proceedings of ITC **19**: 909–918.

[53] Denteneer, D., J.S.H. van Leeuwaarden, and I.J.B.F. Adan (2007). The acquisition queue. *Queueing Systems* **56**(3–4): 229–240.

[54] Denteneer, D. and M.S. Keane (2004). The distribution of success instants in contention trees. *National Laboratory Technical Note* TN-2004/00218.

[55] Denteneer, D. and V. Pronk (2001). On the number of contenders in a contention tree, In: *Proceedings of ITC Specialist Seminar* **14**, Girona: 105–112.

[56] Van Driel, C.-J., P.A.M. van Grinsven, V. Pronk, and W.A.M. Snijders (1997). The (r)evolution of access networks for the information super-highway. *IEEE Communications Magazine* **35**: 2–10.

[57] Dutta-Roy, A. (1999). Cable, it's not just for tv. *IEEE Spectrum* **35**: 53–59.

[58] Data-Over-Cable Service Interface Specifications, Cable Television Laboratories Inc., Public Report SP-RFlv1.1-Pl01-990226.

[59] Digital Video Broadcasting (DVB); interaction channel for Cable TV distribution systems (CATV), working draft (Version 3), June 28, 2000, based on European Telecommunications Standard 300 800 (March 1998).

[60] Evgrafov, M.A. (1966). *Analytic Functions*, Dover, New York.

[61] Fayolle, G. and R. Iasnogorodski (1979). Two coupled processors: The reduction to a Riemann-Hilbert problem. *Zeitschrift für Wahrscheinlichkeitstheorie und Verwandte Gebiete* **47**: 325–351.

[62] Fayolle, G., R. Iasnogorodski, and V. Malyshev (1999). *Random Walks in the Quarter Plane*, Springer, New York.

[63] Fayolle, G., P.J.B. King, and I. Mitrani (1982). The solution of certain two-dimensional Markov models. *Advances in Applied Probability* **14**: 295–308.

[64] Feller, W. (1968). *An Introduction to Probability Theory and its Applications*, Vol. I (3rd edition), Wiley, New York.

[65] Fendick, K.W. and M.A. Rodrigues (1994). Asymptotic analysis of adaptive rate control for diverse sources with delayed feedback. *IEEE Transactions on Information Theory* **40**: 2008–2025.

[66] Feng, W., M. Kowada, and K. Adachi (1998). A two-queue model with Bernoulli service schedule and switching times. *Queueing Systems* **30**: 405–434.

[67] Finkenzeller, K. and R. Waddington (2000). *RFID-handbook*, Wiley, New York.

[68] Franx, G.J. (2001). A simple solution for the M/D/c waiting time distribution. *Operations Research Letters* **29**: 221–229.

[69] Gaier, D. (1964). *Konstruktive Methoden der Konformen Abbildung*, Springer, Berlin.

[70] Gail, H.R., S.L. Hantler, and B.A. Taylor (1996). Spectral analysis of M/G/1 and G/M/1 type Markov chains. *Advances in Applied Probability* **28**, 114–165.

[71] Gakhov, F.D., E.I. Zverovich, and S.G. Samko (1973). Increment of the argument, logarithmic residue and a generalized principle of the argument (in Russian), *Doklady Akademii Nauk SSSR* **213**: 1233–1236 (translated in *Soviet Mathematics Doklady* **14**: 1856–1860).

[72] Gallager, R.G. (1978). Conflict resolution in random access broadcast networks. In: *Proceedings of AFOSR Workshop on Communication Theory and Applications*: 74–76.

[73] Golmie, N., S. Masson, G. Pieris, and D.H. Su (1997). A MAC protocol for HFC networks: Design issues and performance evaluation. *Computer Communications* **20**: 1042–1050.

[74] Golmie, N., Y. Santillan, and D.H. Su (1999). A review of contention resolution algorithms for the IEEE 802.14 networks. *IEEE Communication Surveys and Tutorials* **2**: 2–12.

[75] Gromoll, H.C., A.L. Puha, and R.J. Williams (2002). The fluid limit of a heavily loaded processor sharing queue. *Annals of Applied Probability* **12**: 797–859.

[76] Gyori, I. and G. Ladas (1991). *Oscillation Theory of Delay Differential Equations, with Applications*, Clarendon Press, Oxford.

[77] Hayes, J.F. (1984). *Computer Communication Networks*, Plenum, New York.

[78] Hekstra-Nowacka, E., V. Pronk, L. Tolhuizen, and D. Denteneer (1999). Bandwidth allocation in HFC networks. In: W. Verhaegh, J. Korst, and E. Aarts (eds.) *Proceedings of Scharm'99: Philips workshop on scheduling and resource management*: 129–138.

[79] Henrici, P. (1974). *Applied and Computational Complex Analysis, Vol. I*, Wiley, New York.

[80] IEEE project 802.14. *Cable-TV access method and physical layer specification*. Draft 2 Revision 2.

[81] Iglehart, D.L. (1965). Limit diffusion approximations for the many server queue and the repairman problem. *Journal of Applied Probability* **2**: 429–441.

[82] Iglehart, D.L. and W. Whitt (1970). Multiple channels in heavy traffic I. *Advances in Applied Probability* **2**: 150–177.

[83] Ivanovich, M., M. Zukerman, and R.G. Addie (1997). Performance investigation into an IEEE 802.14 MAC protocol for HFC networks. In: *Proceedings of ICC* 97, Montreal, **2**: 999–1003.

[84] Ivanovich, M. and M. Zukerman (1998). Evaluation and scheduling schemes for an IEEE 802.14 MAC protocol loaded by real traffic. In: *Proceedings of INFOCOM* 98, San Francisco, CA, **3**: 1384–1391.

[85] Jacquet, P., P. Mühlethaler, and P. Robert (2001). Framing protocols on upstream channel in CATV networks: Asymptotic average delay analysis. *INRIA Report 4114*, Rocquencourt.

[86] Janssen, A.J.E.M. and M.J.M. de Jong (2000). Analysis of contention tree-algorithms. *IEEE Transactions on Information Theory* **46**: 2163–2172.

[87] Janssen, A.J.E.M. and J.S.H. van Leeuwaarden (2005). A discrete queue, Fourier sampling on Szegö curves, and Spitzer's formula. *International Journal of Wavelets, Multiresolution and Information Processing* **3**: 361–387.

[88] Janssen, A.J.E.M. and J.S.H. van Leeuwaarden (2005). Analytic computation schemes for the discrete-time bulk service queue, *Queueing Systems* **50**: 141–163.

[89] Janssen, A.J.E.M. and J.S.H. van Leeuwaarden (2005). Relaxation time for the discrete D/G/1 queue. *Queueing Systems* **50**: 53–80.

[90] Johnson, N.L., S. Kotz, and A.W. Kemp (1992). *Univariate Discrete Distributions* (2nd edition), Wiley, New York.

[91] Johari, R. and D.K.H. Tan (2001). A new feedback congestion control policy for long propagation delays. *IEEE/ACM Transactions on Networking* **9**: 818–832.

[92] Kang, K. and B. Steyaert (1999). Bounds analysis for WRR scheduling in a statistical multiplexer with bursty sources. *Telecommunications Systems* **12**: 123–147.

[93] Kaplan, M.A. and E. Gulko (1985). Analytic properties of multiple access trees. *IEEE Transactions on Information Theory* **31**: 255–263.

[94] Kemperman, J.H.B. (1961). *The Passage Problem for a Stationary Markov Chain*, The University of Chicago Press, Chicago.

[95] Kingman, J.F.C. (1970). Inequalities in the theory of queues. *Journal of the Royal Statistical Society, Series B* **32**: 102–110.

[96] De Klein, S.J. (1988). *Fredholm Integral Equations in Queueing Analysis*, Ph.D. thesis, Utrecht University.

[97] Kleinrock, L. (1968). Certain analytical results for the time-shared processors. In: *Proceedings of IFIP68*: 838–845.

[98] Kleinrock, L. (1975). *Queueing Systems Vol. I: Theory*, Wiley, New York.

[99] Kleinrock, L. (1976). *Queueing Systems Vol. II: Computer Applications*, Wiley, New York.

[100] Klimenok, V. (2001). On the modification of Rouché's theorem for the queueing theory problems. *Queueing Systems* **38**: 431–434.

[101] Kobayashi, H. (1978). *Modeling and Analysis. An Introduction to System Performance Evaluation Methodology*, Addison-Wesley, Reading, Massachusetts.

[102] Konheim, A.G. (1975). An elementary solution of the queueing system G/G/1. *SIAM Journal on Computing* **4**: 540–545.

[103] Krichagina, E.V. and A.A. Puhalskii (1997). A heavy traffic analysis of a closed queueing system with a GI/∞ service center. *Queueing Systems* **25**: 235–280.

[104] Kwaaitaal, J.J.B. (1999). *A multi-standard simulation platform for hybrid fiber/coax networks*, Graduate Report, Eindhoven University of Technology, Dept. of Electrical Engineering.

[105] Laevens K. and H. Bruneel (1998). Discrete-time multiserver queues with priorities. *Performance Evaluation* **33**: 249–275.

[106] LAN MAN Standards Committee of the IEEE Computer Society (1999). *Information Technology – Telecommunications and Information Exchange between Systems – Local and Metropolitan Area Networks – Specific Requirements. Part 11: Wireless LAN Medium Access Control (MAC) and Physical Layer (PHY) Specifications* (ISO/IE 8802-11:1999(E); ANSI/IEEE Std 802.11, 1999 Edition, USA).

[107] Law, C., K. Lee, and K.-Y. Siu (2000). Efficient memoryless protocol for tag identification. In: *Proceedings of the 4th International Workshop on Discrete Algorithms and Methods for Mobile Computing and Communications*: 75–84.

[108] Van Leeuwaarden, J.S.H. (2002). *Bandwidth Allocation in HFC Networks using Frame Based Scheduling Strategies*. Final report of the postgraduate program Mathematics for Industry, Stan Ackermans Institute, Eindhoven.

[109] Van Leeuwaarden, J.S.H. (2005). *Queueing Models for Cable Access Networks*, Ph.D. thesis, Eindhoven University of Technology.

[110] Van Leeuwaarden, J.S.H. (2006). Delay analysis for the fixed-cycle traffic light queue. *Transportation Science* **40**: 189–199.

[111] Van Leeuwaarden, J.S.H., T.J.J. Denteneer, and J.A.C. Resing (2006). A discrete-time queueing model with periodically scheduled arrival and departure slots. *Performance Evaluation* **63**: 278–294.

[112] Van Leeuwaarden, J.S.H. and J.A.C. Resing (2005). A tandem queue with coupled processors: Computational issues. *Queueing Systems* **50**: 29–52.

[113] Van Leeuwaarden, J.S.H. and J.A.C. Resing (2004). A two-station network with processor sharing. Unpublished report.

[114] Lindley, D.V. (1952). The theory of queues with a single server. *Proceedings of the Cambridge Philosophical Society* **48**: 277–289.

[115] Malyshev, V. (1972). An analytic method in the theory of two-dimensional random walks. *Siberian Mathematical Journal* **13**: 1314–1329.

[116] Mandelbaum, A. and W.A. Massey (1995). Strong approximations for time-dependent queues. *Mathematics of Operations Research* **20**: 33–64.

[117] Mandelbaum, A., W. Massey, and M. Reiman (1998). Strong approximations for Markovian service networks. *Queueing Systems* **30**: 149–201.

[118] Mandelbaum, A. and G. Pats (1995). State dependent queues: Approximations and applications. In: F.P. Kelly and R.J. Williams (eds.) *Stochastic Networks*, IMA volumes in Mathematics and its Applications **71**: 239–282.

[119] Massey, J.L. (1981). *Collision resolution algorithms and random access communications*. In: G. Longo (ed.) *Multi-user communication systems*, CISM Course and Lecture Notes **265**, Springer, New York.

[120] Mathys, P. and Ph. Flajolet (1985). Q-ary collision resolution algorithms in random-access systems with free or blocked channel access. *IEEE Transactions on Information Theory* **31**: 217–243.

[121] McNeill, D.R. (1968). A solution to the fixed-cycle traffic light problem for compound Poisson arrivals. *Journal of Applied Probability* **5**: 624–635.

[122] Metcalfe, R.M. and D.R. Boggs (1976). Ethernet: Distributed packet switching for local computer networks. *Communications of ACM* **19**: 395–404.

[123] Mikou, N. (1988). A two-node Jackson's network subject to breakdowns. *Communications in Statistics-Stochastic Models* **4**: 523–552.

[124] Mikou, N., O. Idrissi-Kacimi, and S. Saadi (1995). Two processes interacting only during breakdown: The case where the load is not lost. *Queueing Systems* **19**: 301–317.

[125] Miller, A.J. (1963). Settings for fixed-cycle traffic signals. *Operational Research Quarterly* **14**: 373–386.

[126] Mitra, D. (1981). Waiting time distributions for closed queueing network models of shared-processor systems. In: F.J. Kylstra (ed.), *Proceedings of Performance '81*, North Holland Publishing Company, Amsterdam: 113–131.

[127] Mosely, J. and P.A. Humblet (1985). A class of efficient contention resolution algorithms for multiple access channels. *IEEE Transactions on Communication* **33**: 145–151.

[128] Murata, M. and H. Miyahara (1991). An analytic solution of the waiting time distribution for the discrete-time G/G/1 queue. *Performance Evaluation* **13**: 87–95.

[129] Muskhelishvili, N.I. (1992). *Singular Integral Equations.* Dover, New York.

[130] Nauta, H. (1988). *Ergodicity Conditions for a Class of Two-dimensional Queueing Problems*, Ph.D. thesis, Utrecht University.

[131] Neagoe, V.E. (1996). Inversion of the Van der Monde matrix. *IEEE Signal Processing Letters* **3**: 119–120.

[132] Neininger, R. and L. Rüschendorf (2004). A general limit theorem for recursive algorithms and combinatorial structures. *Annals of Applied Probability* **14**: 378–418.

[133] Neuts, M.F. (1989). *Structured Stochastic Matrices of M/G/1 type and Their Applications*, Dekker, Basel.

[134] Newell, G.F. (1960). Queues for a fixed-cycle traffic light. *Annals of Mathematical Statistics* **31**: 589–597.

[135] Palmowski, Z. and S. Schlegel (2002). Modeling of cable access networks. *National Laboratory Technical Note* TN-2002/820.

[136] Palmowski, Z., S. Schlegel, and O.J. Boxma (2003). A tandem queue with a gate. *Queueing Systems* **43**: 349–364.

[137] Pollaczek, F. (1932). Zur Theorie des Wartens vor Schaltergruppen. *Elektronische Nachrichtentechnik* **9**: 434–454.

[138] Pollaczek, F. (1952). Fonctions charactéristiques de certaines répartitions définies au moyen de la notion d'ordre, Application à la théorie des attentes. *Comptes rendus de l'Académie des sciences-Paris* **234**: 2334–2336.

[139] Pólya, G. and G. Szegö (1972). *Problems and Theorems in Analysis, Vol. I*, Springer, New York.

[140] Pólya, G. and G. Szegö (1976). *Problems and Theorems in Analysis, Volume II*, Springer, New York.

[141] Polyzos, G.C. and M. Molle (1994). A queueing theoretic approach to the delay analysis for the FCFS 0.487 conflict resolution algorithm. *IEEE Transactions on Information Theory* **39**: 1887–1906.

[142] Powell, W.B. (1985). Analysis of vehicle holding and cancellation strategies in bulk arrival, bulk service queues. *Transportation Science* **19**: 352–377.

[143] Powell W.B. and P. Humblet (1986). The bulk service queue with a general control strategy: Theoretical analysis and a new computational procedure. *Operations Research* **34**: 267–275.

[144] Pronk, S. and M. de Jong (1998). Multi-standard simulation platform for hybrid fiber/coax networks; I Standard survey, II Basic architecture. *National Laboratory Technical Note* TN 179/98.

[145] Pronk, S.P.P., E. Hekstra-Nowacka, L. Tolhuizen, and D. Denteneer (1999). Description and performance analysis of the DVB/DAVIC MAC protocol for HFC Networks. *National Laboratory Report* 7105.

[146] Prabhu, N.U. (1980). *Stochastic Storage Processes. Queues, Insurance Risk, and Dams*, Springer, New York.

[147] Regterschot, G.J.K. (1987). *Wiener-Hopf Factorization Techniques in Queueing Models*, Ph.D. thesis, University of Twente.

[148] Resing, J.A.C. and L. Örmeci (2003). A tandem queueing model with coupled processors. *Operations Research Letters* **31**: 383–389.

[149] Roberts, L.G. (1972). *Aloha Packet System with and without Slots and Capture* (ASS Note 8). Stanford, CA: Stanford Research Institute, Advanced Research Projects Agency, Network Information Center.

[150] Sala, D., J. Limb, and S. Khaunte (1998). Adaptive control mechanisms for cable modem MAC protocols. In: *Proceedings of INFOCOM 98*, **3**, San Francisco, CA: 1392–1399.

[151] Scherr, A.A. (1967). *An analysis of time-shared computer systems*, MIT Press, Cambridge, MA.

[152] Servi, L.D. (1986). D/G/1 queues with vacations. *Operations Research* **34**: 619–629.

[153] Sevcik, K.C. and I. Mitrani (1979). The distribution of queueing network states at input and output instants, In: M. Arato *et al.* (eds.) *Proceedings of Performance* 79, North Holland Publishing, Amsterdam: 319–335.

[154] De Smit, J.H.A. (1995). Explicit Wiener-Hopf factorizations for the analysis of multidimensional queues. In: J.H. Dshalalow (ed.) *Advances in Queueing. Theory, Methods and Open Problems*, CRC, Boca Raton: 293–310.

[155] Smith, W.L. (1953). On the distribution of queueing times. *Proceedings of the Cambridge Philosophical Society* **49**: 449–461.

[156] Spitzer, F.L. (1956). A combinatorial lemma and its application to probability theory. *Transactions of American Mathematical Society* **82**: 323–339.

[157] Stadje, W. (1997). A new approach to the Lindley recursion. *Statistics & Probability Letters* **31**: 169–175.

[158] Syski, R. (1967). Pollaczek's methods in queueing theory. In: R. Cruon (ed.) *Queueing Theory, Recent Developments and Applications*, Oxford Univerity Press, London: 33–60.

[159] Takács, L. (1962). *Introduction to the Theory of Queues*, Oxford University Press, Oxford.

[160] Tanenbaum, A.S. (1981). *Computer Networks*, Prentice-Hall, Englewood Cliffs.

[161] Tijms, H.C. (1994). *Stochastic Models: An Algorithmic Approach*, Wiley, New York.

[162] Titchmarsh, E.C. (1939). *The Theory of Functions* (2nd edition), Oxford University Press, New York.

[163] Tsybakov, B.S. (1985). Survey of USSR contributions to random multiple-access communications. *IEEE Transactions on Information Theory* **31**: 143–165.

[164] Tsybakov, B.S. and V.A. Mikhailov (1978). Free synchronous packet access in a broadcast channel with feedback. *Problemy Peredachi Informatsii* **14**: 32–59.

[165] Tsybakov, B.S. and V.A. Mikhailov (1980). Random multiple access of packets: Part and try algorithm. *Problemy Peredachi Informatsii* **16**: 305–317.

[166] US 7,251,251 (2007). Method and system for transmitting a plurality of messages. United States Patent.

[167] Van Uitert, M.J.G. (2003). *Generalized Processor Sharing Queues*, Ph.D. thesis, Eindhoven University of Technology.

[168] Vaulot, E. (1951). Les formules d'Erlang et leur calcul pratique. *Annals of Telecommunications* **6**: 279–286.

[169] Webster, F.V. (1958). Traffic signal settings. *Road Research Laboratory Technical Report* No. **39**, HMSO, London.

[170] Whitt, W. (1980). Some useful functions for functional limit theorems. *Mathematics of Operations Research* **5**: 67–85.

[171] Whitt, W. (1984). Heavy-traffic approximations for service systems with blocking. *AT&T Bell Laboratories Technical Journal* **63**: 689–708.

[172] Whitt, W. (2002). *Stochastic Process Limits*, Springer, New York.

[173] Winands, E.M.M., D. Denteneer, J.A.C. Resing, and R. Rietman (2003). A finite-source feedback queueing network as a model for the IEEE 802.11 distributed coordination function. *SPOR-Report 2003-22*. Dept. of Mathematics and Computer Science, Eindhoven University of Technology.

[174] Winands, E.M.M., T.J.J. Denteneer, J.A.C. Resing, and R. Rietman (2005). A finite-source queueing model for the IEEE 802.11 DCF. *European Transactions on Telecommunications* **16**: 77–89.

[175] Yin, W.-M. and Y.-D. Lin (2000). Statistically optimized minislot allocation for initial and collision resolution in hybrid fiber coaxial networks. *IEEE Journal on Selected Areas in Communication* **18**: 1764–1773.

[176] Zhao, Y.Q. and L.L. Campbell (1996). Equilibrium probability calculations for a discrete-time bulk queue model. *Queueing Systems* **22**: 189–198.

About the Authors

Dee Denteneer was born in Brunssum, The Netherlands, in 1960. He studied mathematics and computer science (Utrecht, 1984) and obtained a PhD degree in mathematics (Eindhoven, 2005) supervised by Onno Boxma, Sem Borst, and Michael Keane.

Dee started his career at the University of Utrecht as a scientific programmer in the project 'Analysis of discrete data with small numbers of observations' under the supervision of Albert Verbeek. Next, between 1984 and 1988, he worked at the Dutch Central Statistical Office on the Blaise language for questionnaire description. Currently, he is a principal research scientist at Philips Research in Eindhoven. His main interest is in statistics and applied probability and their application in industrial research areas such as MPEG encoding, speech recognition, secure biometrics, and data transmission systems. Since 2006 he has been working on the performance analysis of wireless mesh networks, and their standardisation in IEEE 802.11s.

Dee is married to Marion Hermans, and they have four children, Thijs, Ellen, Anne, and Paul.

Johan van Leeuwaarden was born in Eindhoven, The Netherlands, in 1978. He studied econometrics (Tilburg, 2000) and mathematics (Eindhoven, 2002). Supervised by Jacques Resing, Onno Boxma and Sem Borst, he obtained a PhD degree in mathematics (with honors, Eindhoven, 2005). His dissertation received the Beta PhD Award (2006) and the INFORMS Telecommunications Dissertation Award (2008).

Johan currently works as an assistant professor in the stochastic operations research group of the faculty of mathematics and computer science in Eindhoven. He is funded by a VENI grant of the Netherlands Organization for Scientific Research (NWO). He is also a research fellow of EURANDOM. His interests lie in applied probability, queueing theory and analysis, with possible application to the performance analysis of computer systems and communication networks.